TABLE OF CONTENTS

ABOUT CHEMISTRY

Chemistry is a fascinating field of science. It is unique from other fields of science - like biology, astronomy, etc - because it is an analysis of the **invisibility** in the universe. In the world of chemistry, scientists cannot visibly see their experiments in a macroscopic level. Instead, they need to focus on the molecular level.

Chemistry is important to fully understand because it provides us with the answers to why the world proceeds the way it currently does.

Chemistry can help us understand why reactions occur.

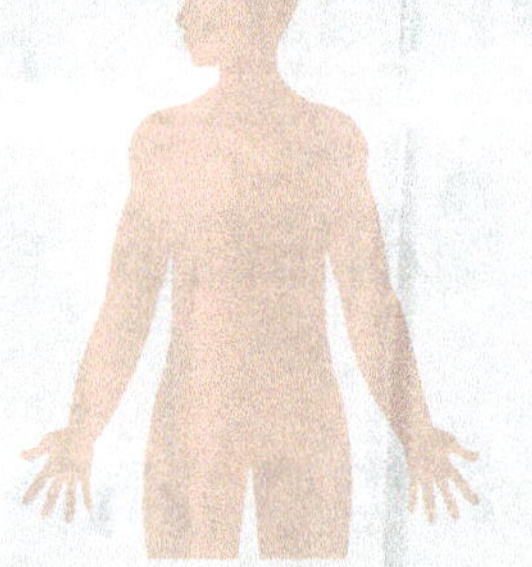

Chemistry can help us understand how our bodies react.

Chemistry can help us understand the nature of the universe.

01 THE ATOM

An **atom** is a particle with a **positive nucleus** in a **cloud of negative electrons**. The nucleus has **positively-charged protons** and **neutrally-charged neutrons**, which make the overall charge **positive**. This is useful for various reactions because the positive protons of one atom's nucleus can **attract** the negative electrons of another atom.

An atom makes up everything on Earth. From the tiniest microorganism to a human being, atoms compose all life. The ways that atoms interact with other atoms, molecules, and compounds determine how certain reactions occur in the universe.

Aside from atoms, **elements** exist as well. An **element** is a chemical substance that cannot be broken down into simpler substances. Some examples are **carbon**, as shown above, **oxygen**, **sodium**, **hydrogen**, etc. All of these elements are listed on a table known as the **Periodic Table of Elements**.

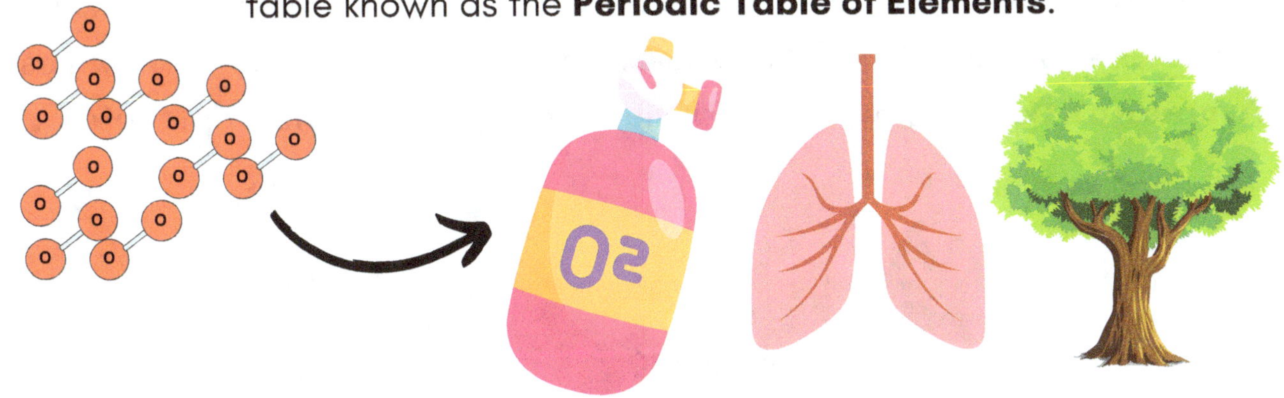

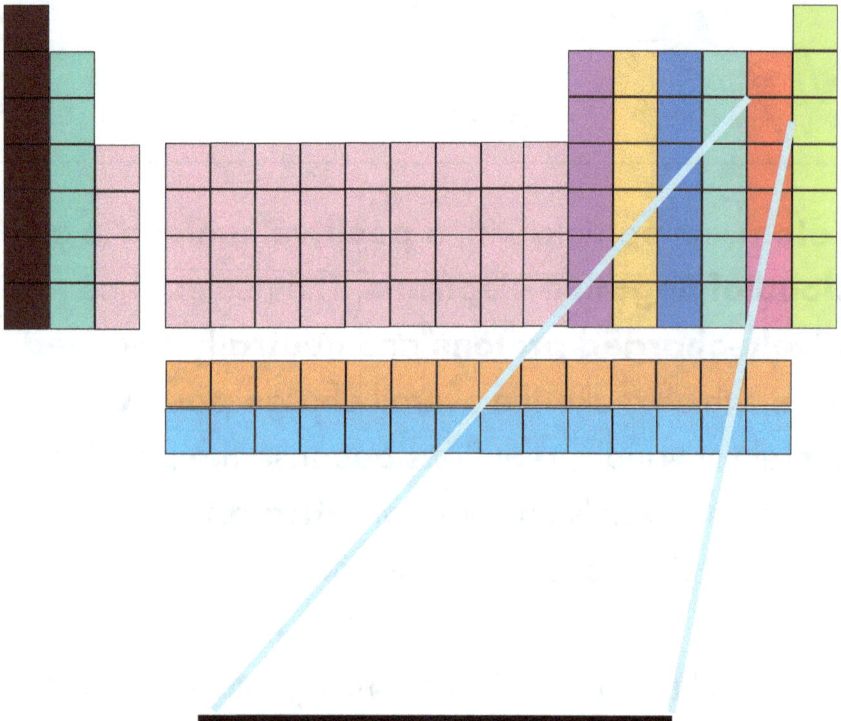

The Periodic Table of Elements is arranged according to **increasing atomic number**. The **atomic number** is the number of **protons** and **electrons** - and for some elements, **neutrons** - an element has. The second number is the **atomic mass**, which is often referred to as the **weighted average atomic mass** of the number of protons and neutrons in the nucleus of an atom.

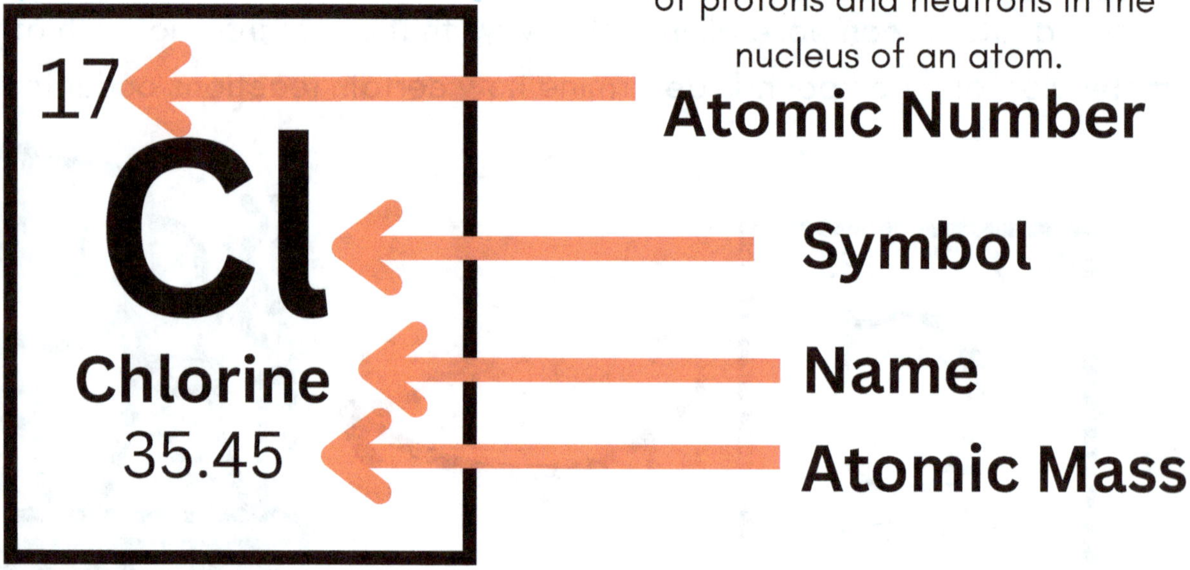

17

Cl

Chlorine

35.45

Atomic Number

Symbol

Name

Atomic Mass

Atomic masses on the periodic table show how many grams are present in one **mole** of an element. The **atomic mass unit (amu)** is the standard unit of atomic mass. A **mole** of a substance is equal to **6.02×10^{23}** of something, whether that may be **grams, particles, molecules, atoms, etc**. For example, **one mole** of water molecules is equal to **6.02×10^{23}** molecules. This number, **6.02×10^{23}** is known as **Avogadro's Number**, a concept similar to a **dozen, couple, triple, etc.**

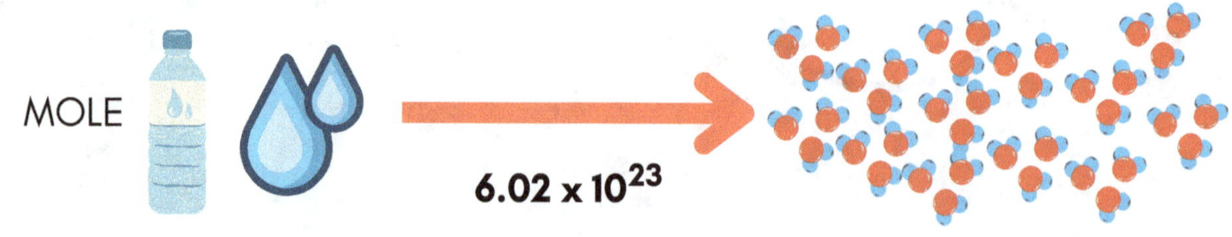

The **molar mass** of a chemical substance basically states how many **grams** is in a **mole** of a substance. The molar mass of an element is represented in **moles** and can be converted into **atoms** via **Avogadro's number**. In nature, because of **isotopes**, a form of an element that has the **same number of protons but different number of neutrons**, different samples of the same chemical substance can have **varying** molar masses. However, the **molar mass** gives you the **most common** mass.

Demonstration (calculating average atomic mass):

Magnesium (Mg)

Mg-24: 78.99% abundant
Mg-25: 10% abundant
Mg-26: 11.01% abundant

Average Atomic Mass:

$(0.7899 \times 24) + (0.10 \times 25) + (0.1101 \times 26)$

$= 24.3$ g/mol $= 24.3$ amu

The **molecular mass** is essentially the molar mass of a molecule. You can find the molecular mass by **adding** all of the individual masses of the **elements** that the molecule is composed of.

Demonstration (calculating molecular mass):

Glucose ($C_6H_{12}O_6$)

Molecular Mass:

$(6 \times 12.01) + (12 \times 1.008) + (6 \times 16.00)$

$= 180.16$ g/mol $= 180.16$ amu

Believe it or not, with **molar mass, Avogadro's number, molecular mass, etc**, we can **convert** between units in chemistry. This is useful for **laboratory experiments** because it can help you decide how much of a particular substance to use. This "conversion" between units is known as **stoichiometry**.

Demonstration (stoichiometry!):

How many glucose molecules are in one kilogram of glucose?

1 kg = 1,000 g
Molecular Mass: 180.16 g/mol
Avogadro's Number: 1 mol = 6.02 x 10^{23} molecules

1,000 g glucose / (180.16 g glucose/1 mol glucose)
$= 5.55$ mol glucose

5.55 mol glucose x (6.02 x 10 molecules/1 mol
$= 3.34 \times 10^{24}$ molecules glucose

Molarity is one of the other units in chemistry. It is expressed as the **number of moles of solute** (substance that you mix into the **solution**) over the **liters of solution**. In short, it is expressed as either **M** for molarity or **mol/L**. It is relatively simple to determine: you just find the number of moles of solute and divide it by the **volume** of solution expressed in liters. An **important fact** is that the volume does **not** include the solute.

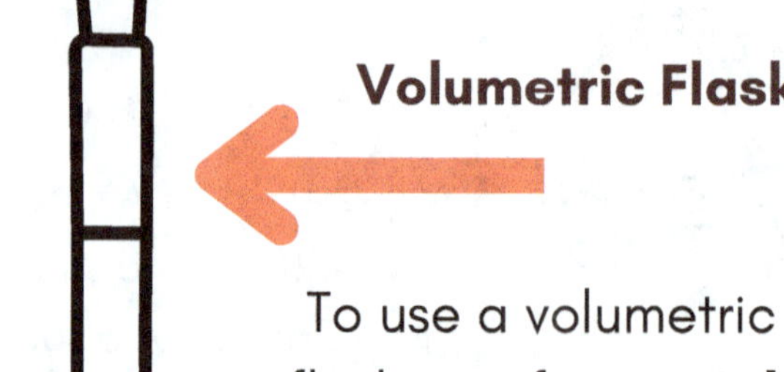

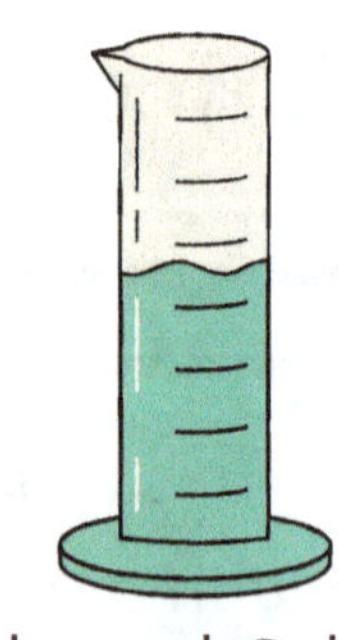

Volumetric Flask

To use a volumetric flask, you first **weigh out** a small amount of **solute** using a scale and dissolve it in the volumetric flask with a little bit of liquid. Then, you fill the flask up with the liquid up to the line.

In a **laboratory setting**, a **volumetric flask** is often the preferred tool to use to make a solution of a specific molarity. There are various types of volumetric flasks, such as **50-mL** and **100-mL**. It is important to convert **mL** into **L** when measuring molarity.

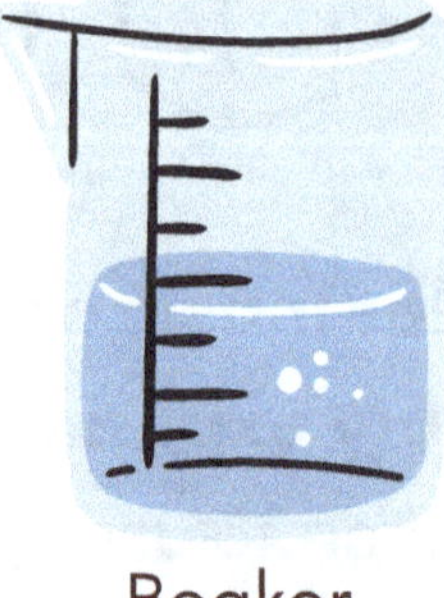

Beaker

Graduated Cylinder

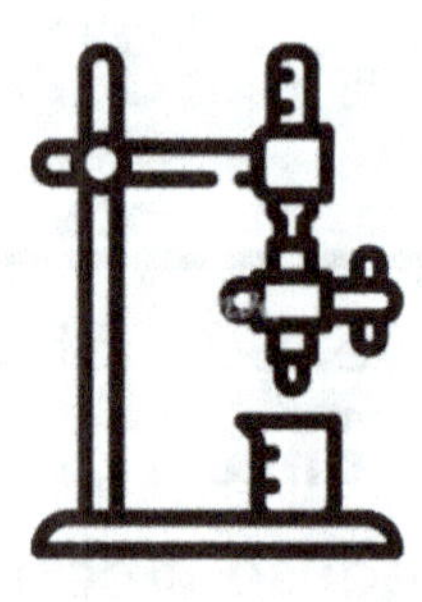

Titration Burette

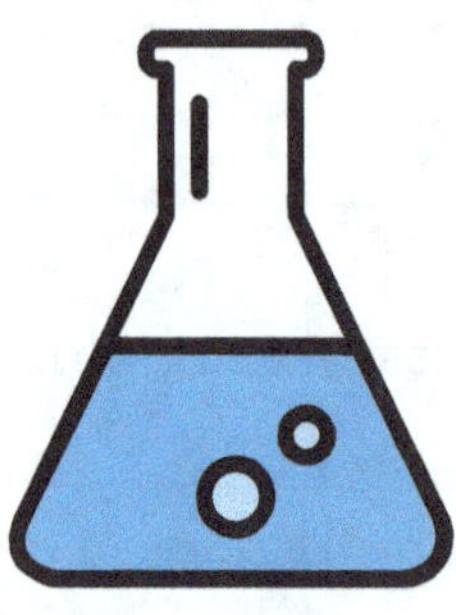

Erlenmeyer Flask

Bunsen Burner

Funnel

02 MOLECULES

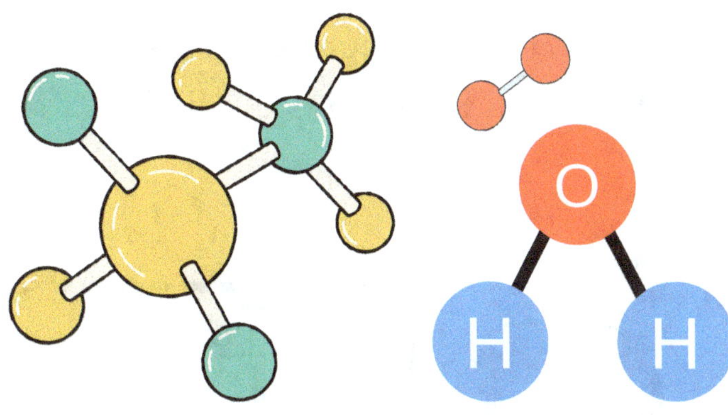

A **molecule** is a combination of **atoms**, and in most cases, a **molecule** has multiple of the **same atom**. However, once you bond **different atoms**, you have a **compound**. Both molecules and compounds react with other substances.

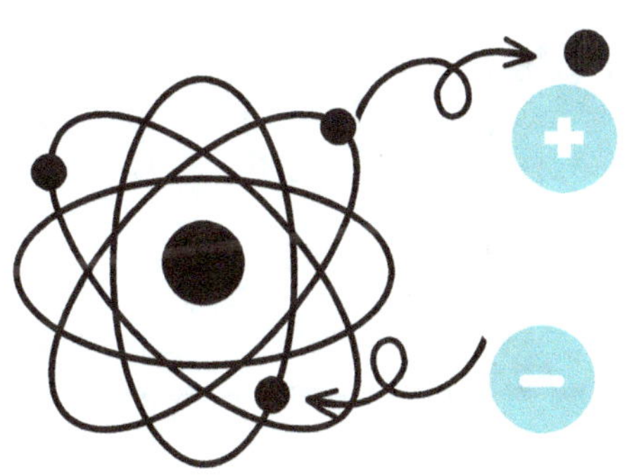

An **ion** is different from an **atom** because an **ion** is what results once an atom **gains** or **loses** an electron. Gaining an electron is equivalent to "**gaining a negative charge**," thus creating a **negative ion (i.e. anion)**. Losing an electron is equivalent to "**losing a negative charge**," thus creating a **positive ion (i.e. cation)**.

When an atom gives electrons to another, a **difference in charges** occurs as the giving atom becomes **positive** and the receiving atom becomes **negative**. This creates an **attraction** between the new **ions**, and this is what fuels an **ionic bond** that occurs in chemistry.

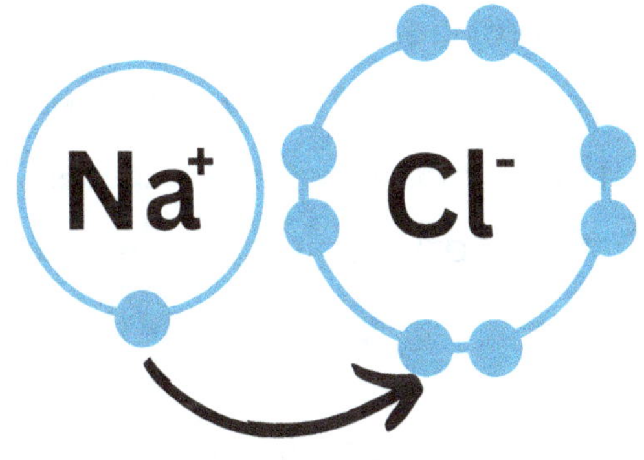

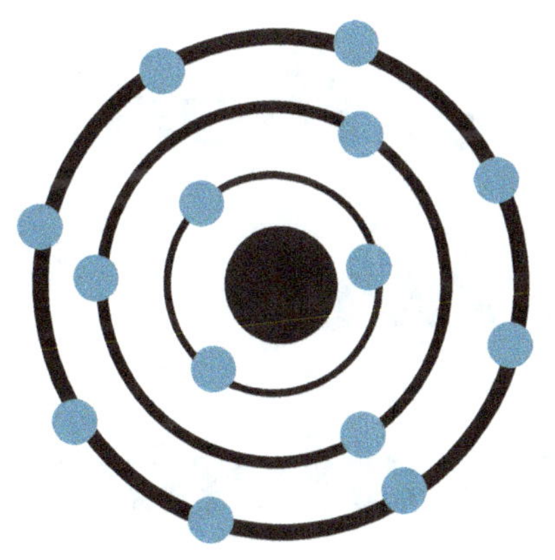

BUT WHY DO ATOMS BOND? WHY DO IONS TRANSFER ELECTRONS? It's because atoms desire **a full valence (outermost) shell**, which means that they want **eight valence (outermost) electrons**. On the periodic table, **the last column of elements (Group 18)** have **ideal gas atoms**, which have **eight valence electrons** and are thus the most stable.

There are **three types** of **formulas** to express molecules and compounds: **empirical, molecular,** and **structural**. The **empirical formula** is the simplest way of representing the different atoms of a compound whereas the **molecular formula** is an accurate representation of the compound's composition. The **structural formula** is a little bit different because it shows a model of the compound.

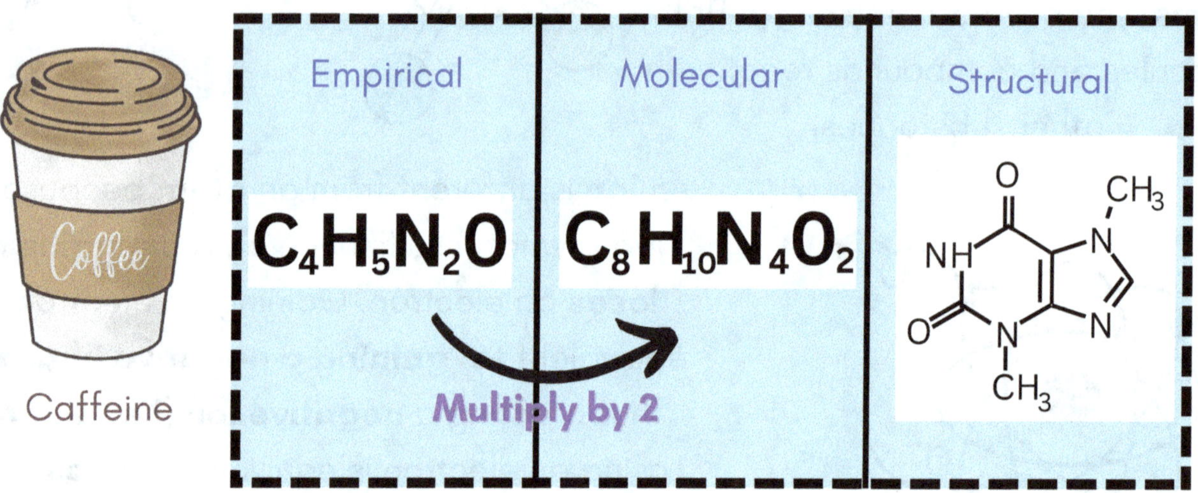

You can express an **empirical formula** in **words**. For example, for **caffeine**, you can say "**for every four carbons, there are five hydrogens**" to express the relationship between **carbon** and **hydrogen**.

An interesting concept is that in some cases, like in a **water molecule**, the **empirical** and **molecular** formula can be the same, simply because the molecular formula can't be simplified.

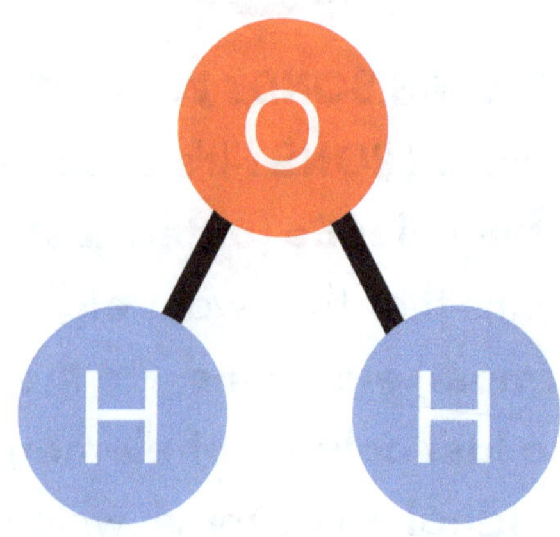

For every oxygen molecule, there are two hydrogen atoms.

The periodic table is expressed in **rows (periods)** and **columns (groups)**. Elements belonging to the same column or group can share **physical** and **chemical properties,** including **standard state (solid, liquid, or gas),** **conductivity,** etc.

The diagram to the right is a **Bohr model** of an atom. This type of model shows the atom with a **nucleus** surrounded by **electrons** in different "rings," which represent **energy levels**. The electrons in the smallest ring are the closest to the nucleus, which means that they are the most attracted to the nucleus as opposed to the electrons in the largest ring.

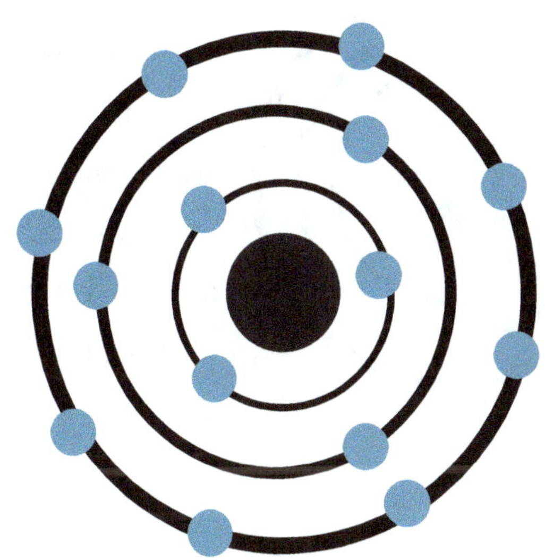

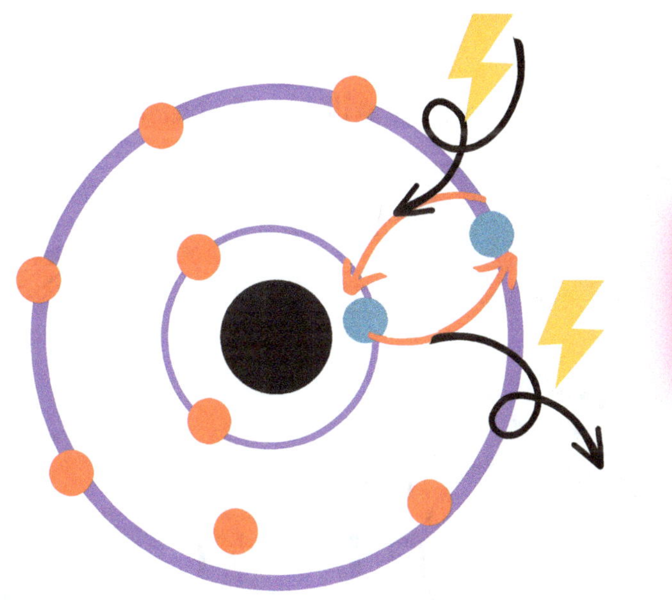

In chemistry, atoms have a **ground state** and an **excited state**. A **ground state** is the **neutral version** of the atom, and this is the **stable, lowest energy** form of the atom. An **excited state** is the **energized version**, and this occurs when electrons **absorb energy** and "jump" to a higher energy level.

Atoms prefer a configuration with the **lowest** energy. Thus, the electron shells of an atom are filled up from the inside out (lowest energy/**1n** → highest energy/**3n**).*1n** can hold two electrons, **2n** can hold eight, and **3n** can hold up to eighteen.*

In chemistry, the **octet rule** is very important and states that atoms will work to create **valence shells**, the **outermost shells**, with **eight electrons**, or however many electrons they need to fill up their valence shells. At this state, the atoms will be the **most stable**. Atoms will either gain or lose electrons to get to this state, which results in many properties in chemistry.

Sodium (Na)

The number of **valence electrons** *increases* from **left to right**. Elements that have **one** or **two valence electrons** are very reactive because they desire to lose their valence electrons to achieve an octet. **Group 18** is the special group, and all elements besides **helium** have **eight valence electrons**. **Helium** has two. This group is known as the **noble gases.**

Names (Most Common Only)!
1. Group 1: **Alkali Metals**
2. Group 17: **Halogens**
3. Group 18: **Noble gases**

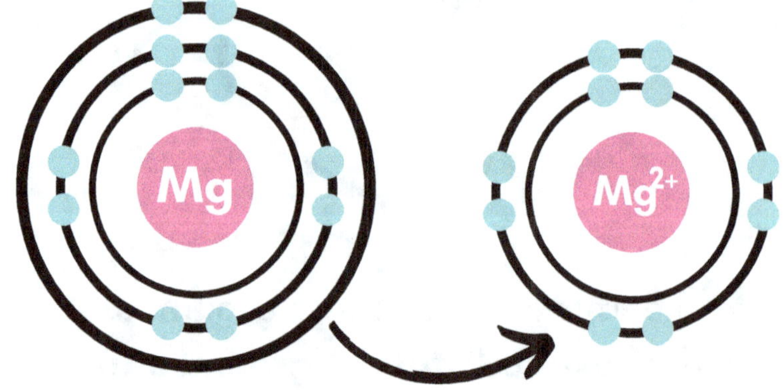

We keep talking about **shells**, which is a very vague term. What are **shells** in chemistry?

Shells are **energy levels** of an electron! Electrons don't really *circle* the nucleus, but rather spend most of their time in specific regions around the nucleus, known as **electron orbitals**. As the energy levels *increase*, the electrons are *further* away from the nucleus. **Sub-shells** are sets of one or more orbitals. They are the result of breaking each electron shell down. They are designated by the letters **s (2)**, **p (6)**, **d (10)**, and **f (14)**, and each letter indicates a **different shape**.

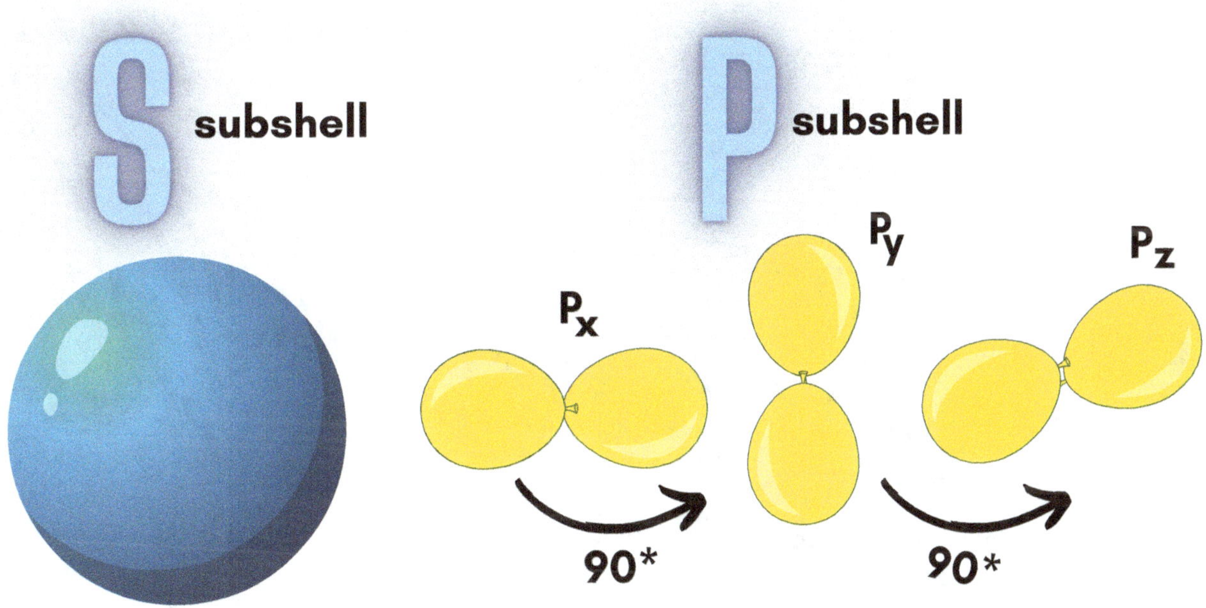

S sub-shells have a **single, spherical orbital. P sub-shells** contain **three dumbbell-shaped orbitals (almost like balloons!)** at right angles to each other. Atoms with many electrons may place some of their electrons in **d** and **f sub-shells**.

The **first electron shell (1n)** fills with electrons first because it is the shell that is the closest to the nucleus. It contains one s-orbital.

The **second electron shell (2n)** contains one s-orbital plus three p-orbitals, each holding two electrons.

The **third electron shell (3n)** contains an s orbital and three p orbitals. Elements in the third row of the periodic table put their electrons in this shell.

All elements and ions have a particular way of organizing their electrons. This method is known as **electron configuration**, which contains information about the **spatial arrangement, number,** and **energy levels** of the electrons. Before we begin writing electron configurations, however, we should stop and consider three important rules.

Aufbau Principle

The orbital with the **lowest energy (1s)** is filled first, followed by higher energy orbitals.

Pauli Exclusion Principle

There are **two** electrons in orbitals, and each electron spins in a different direction (**opposite** directions).

Hund's Rule

The **empty orbitals** are favored over the **occupied orbitals** for electrons.

The **Aufbau Principle** is helpful when drawing the electron configuration of an atom. The **Aufbau "building" diagram** is often used to determine the order of which orbitals are filled up with electrons. After calcium, we will place electrons in the 3d orbital before filling the 4s orbital because in this case, the **4s orbital energy is higher than the 3d orbital energy**.

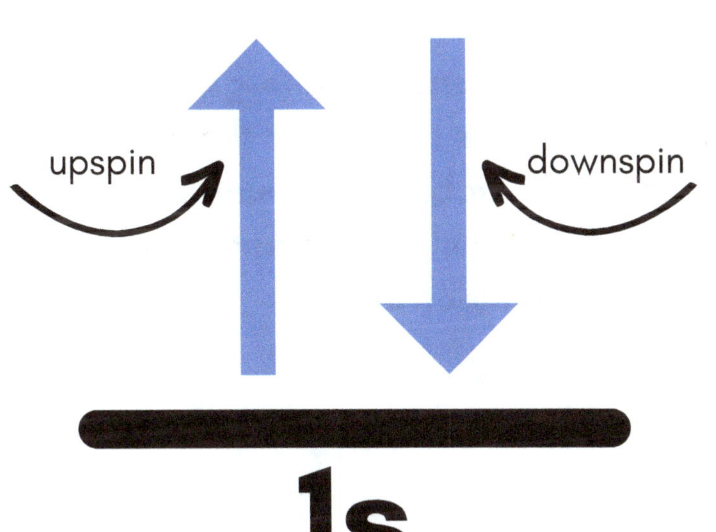

In the **Pauli Exclusion Principle**, the terms **up-spin** and **down-spin** are used, but they mean that electrons spin **clockwise** (up-spin) or **counterclockwise** (down-spin).

To remember **Hund's Rule**, which states that electrons tend to fill **empty orbitals on the same sub-shell** before completing filling orbitals, remember that in a **school bus**, one would be more inclined to sit on an empty seat if there is one as opposed to sharing a seat.

Since we have gotten the three key principles out of the way, let's talk about **writing electron configurations** of different atoms and ions.

Let's look at **carbon**, which has **6 electrons in total** because, according to the periodic table, it has an **atomic number of 6**. The first energy level has n different sub-shells (s), and n different types of orbitals ($1s$). Therefore, the second energy level has **two** different sub-shells (s and p) and **four** different orbitals - **2s, 2p(x), 2p(y), 2p(z)**. Every orbital contains a maximum of **two electrons**. Carbon only requires **three** orbitals, each containing two spots for electrons, since carbon has six electrons.

Carbon's electron configuration is:

$$1s^2 2s^2 2p^2$$

You can tell how many **valence electrons** an atom has by looking at the **greatest nth-orbitals' electrons**. In this case, the **second** ($n = 2$) orbitals are the valence electrons.

You can also write the **electron configurations** in **noble-gas configuration**. It gets tedious to write all of the orbitals out, so the noble-gas configuration takes the configuration of the **noble gas element** that precedes the current element but is closest to the element and adds the orbitals that contain the **valence electrons**.

Helium (He) is the noble gas that **precedes carbon** but is **closest to carbon**, and its electron configuration is just **$1s^2$**. Carbon has "**$1s^2$**" in its electron configuration, so you can rewrite carbon's electron configuration as **[He] $2s^2 2p^2$**.

Let's shift to a different topic: **Coulomb's Law**. Coulomb's Law states that the **magnitude** of the **force** between two charged particles depends on the magnitude of the charges of the two particles and the **distance** between them. The **Coulomb force** is the **attraction** or **repulsion** of particles due to their **electric charges**.

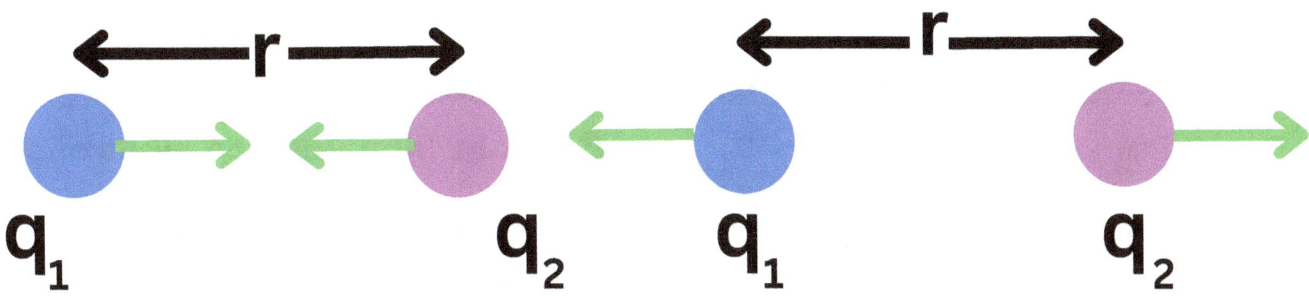

With increasing charges (–2 & +2 vs –3 & +3) and decreasing distance between the two atoms, the strength of the attraction increases, which makes it more difficult to pull the two atoms apart.

NaF vs. MgO

The **ionization energy** refers to how much energy is needed to **take electrons** from an atom. This is different from **electron affinity**, which refers to how much energy is released when we **add** an electron to an atom. If energy is **released** once an electron is added, the electron affinity will be **negative**. If energy is **absorbed** once an electron is added, which means it takes energy to add that electron, the electron affinity will be **positive**.

The **electronegativity** is the measure of the ability of an atom to attract electrons from other atoms or molecules.

The **electron screening** is essentially when a valence electron is attracted to the positive nucleus and moves closer to the nucleus but is blocked by the negative **core electrons (non-valence electrons)**.

The **electronegativity** is the measure of the tendency of an atom to attract electrons from other atoms or molecules.

The **electron screening** is essentially when the negative valence electron is attracted to the positive nucleus and moves closer to the nucleus but is repelled by the negative **core electrons (non-valence electrons)**.

The **atomic radius** is how far the nucleus is from the outermost electrons, which basically means how large the atom is.

The **greater** the number of **protons**, the **smaller** the atom is. The **greater** the number of energy levels, the **larger** the atom is. These concepts explain **periodicity**.

The **ionic radius** is essentially the **atomic radius** of an **ion**. An **ANION's ionic radius** is *larger* than the atom's atomic radius because if an *anion gains an electron*, there will be *more repulsion* from the extra electron, which will **expand** the radius. A **CATION's ionic radius** is *smaller* than the atom's atomic radius because, again, a *cation loses an electron*, meaning there will be *less repulsion* from the removed electron, thereby **shrinking** the radius.

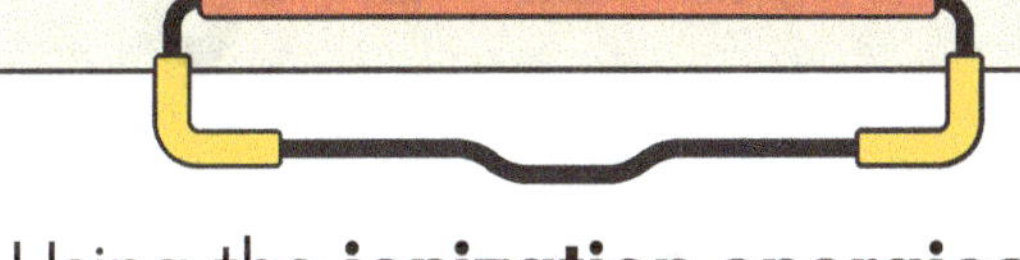

Using the **ionization energies**, you can determine how many valence electrons (and thus group number) of the atom. Say, you have a **lithium atom:** the **first ionization energy** is about **520 kJ/mol**, and the **second ionization energy** is about **7300 kJ/mol**, which is *way larger* than the first. The reason for this dramatic increase is a **decrease** in both **electron shielding** and **distance** from the nucleus in the second valence electron. Therefore, you know that **lithium** has **one valence electron.**

Removing a **core electron** has a **much higher** ionization energy than removing a **valence electron** because the **valence electron** is in the outermost energy level and further away from the nucleus than a **core electron** (non-valence). Therefore, if your **first ionization energy** is **500 kJ, second** is **900 kJ, third** is **1200 kJ,** and **fourth** is **3000 kJ,** you know that the fourth electron is the **first core electron,** meaning that the first **three** electrons were **valence electrons.** This atom would belong to **group 13** and have **three valence electrons.**

We need to know how to **predict** ionic compounds. Remember, the **main goal** of forming ionic compounds is to help **all ions** achieve a **full valence shell.** If an atom has **three or fewer** valence electrons, the atom will most likely **lose them**. If an atom has **five or more** valence electrons, the atom will most likely **gain more electrons**. However, atoms in **group 14** (with **four valence electrons**) can sometimes **lose or gain** electrons, depending on the different electronegativity.

We have **ALUMINUM (with three valence electrons)** and **CHLORINE (with seven valence electrons)**. For aluminum to achieve an octet/full shell, it needs to **give up its three electrons**, and for chlorine to achieve an octet/full shell, it needs to **gain one electron.**

Let's talk about **photoelectron spectroscopy**. It is a way of examining the electron configuration of a certain type of atom. It gives you a sense of **how many electrons** have **specific binding energies.**

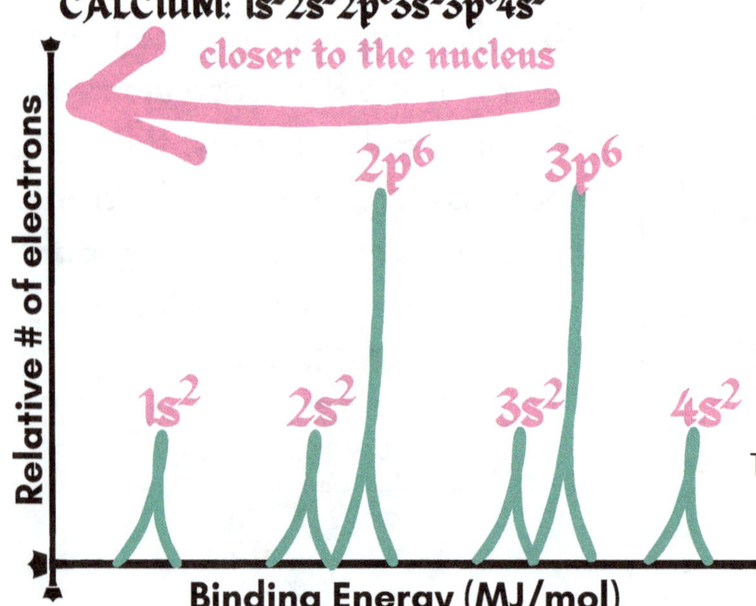

The different heights of the spikes represent the varying number of electrons (either two electrons or six electrons).

For **potassium (K)**, the spikes would shift **slightly to the right** since potassium's electrons would be slightly less attracted to the nucleus, which would have **19 protons** instead of **20**. The **last spike** would **half its size** since instead of **two valence electrons**, potassium only has **one**.

03 INTERACTIONS

An **ionic solid** is held together by the **attraction** between ions, and they occur between **metals** and **nonmetals**. Ionic solids are usually **solids at room temperature** and have **high melting and boiling points**. They **conduct electricity** when dissolved in water because of the **aqueous ions**. However, as a solid, they do not because the electrons "belong" to a particular ion.

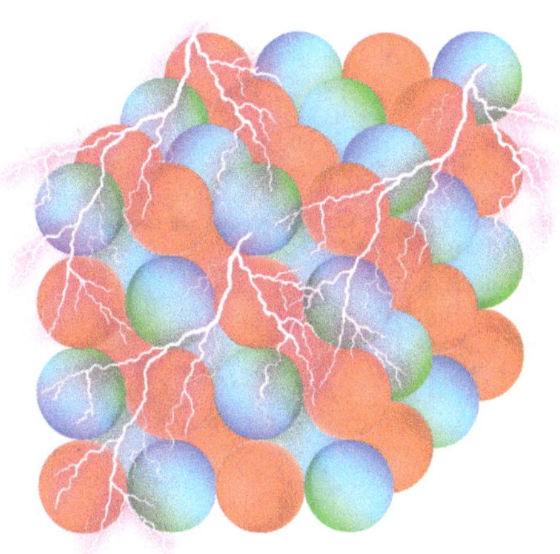

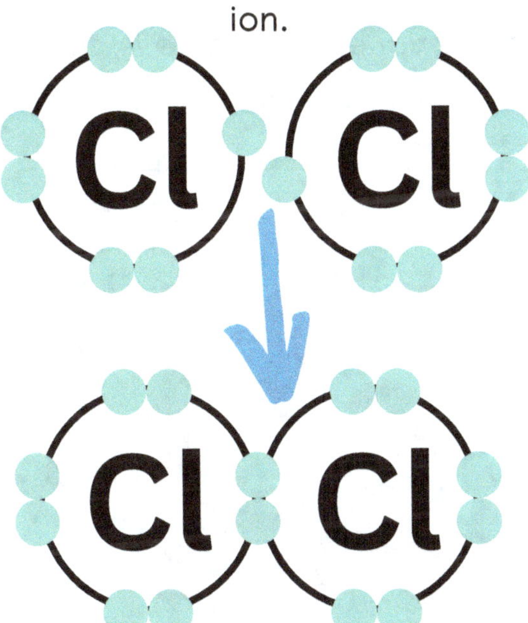

A **covalent bond** forms when two atoms with similar electronegativity share their valence electrons like two friends. In covalent bonds, multiple bonds are stronger than single bonds. The first covalent bond is a **sigma bond;** the rest are **pi bonds**.

A **polar covalent bond** is when electrons are **more attracted** to the **more electronegative atom,** forming **partial positive** and **partial negative charges.**
- **Dipole moments!**

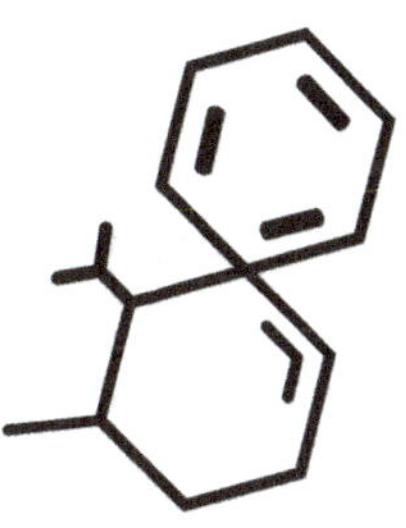

A **nonpolar covalent bond** is when electrons are **shared equally**. This occurs in atoms with **identical electronegativity** like H_2, O_2, F_2, Br_2, I_2, N_2, Cl_2

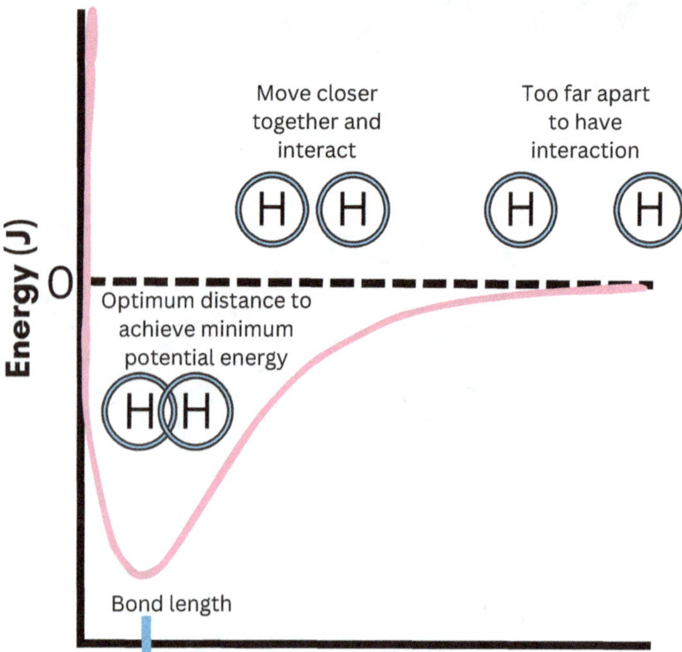

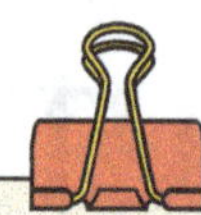

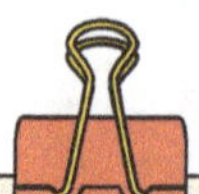

The **internuclear distance** is the distance at which the **minimum potential energy** is obtained. This also determines **bond length**. The minimum potential energy will occur when the **attractive** and **repulsive** forces are balanced. One important thing to keep in mind is that **zero Joules** of energy is **NOT** the minimum potential energy. *In all potential energy cases, the more negative the potential energy value is, the lower the energy is.*

A **network solid** is made up of atoms that are held together in a lattice of covalent bonds. This type of solid is **very hard** and has **high melting** and **boiling points**. Since electrons are not free to move about the lattice, the solid is a **poor conductor of electricity**. Because **carbon** (e.g. **diamond** or **graphite**) and **silicon** (SiO_2 – **quartz**) have **four valence electrons**, they are able to form a large number of covalent bonds and are thus the **most commonly seen network solids**.

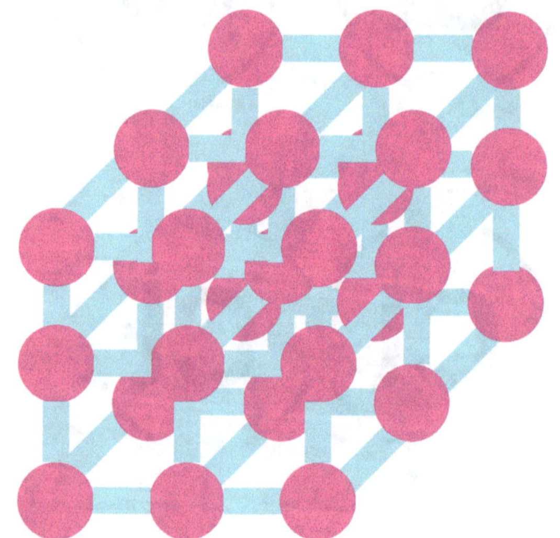

A **metallic solid** is represented by the **"sea of electrons" model**, in which the **positive cores** of the metal atoms are surrounded by **electrons**. This is why metals **conduct electricity** and are **malleable** and **ductile**.

Alloys can be formed from mixtures of metals.
Interstitial alloy: metals with two **different** sizes combine (e.g. **steel**).
Substitutional alloy: metals with **similar** sizes combine (e.g. **brass**).

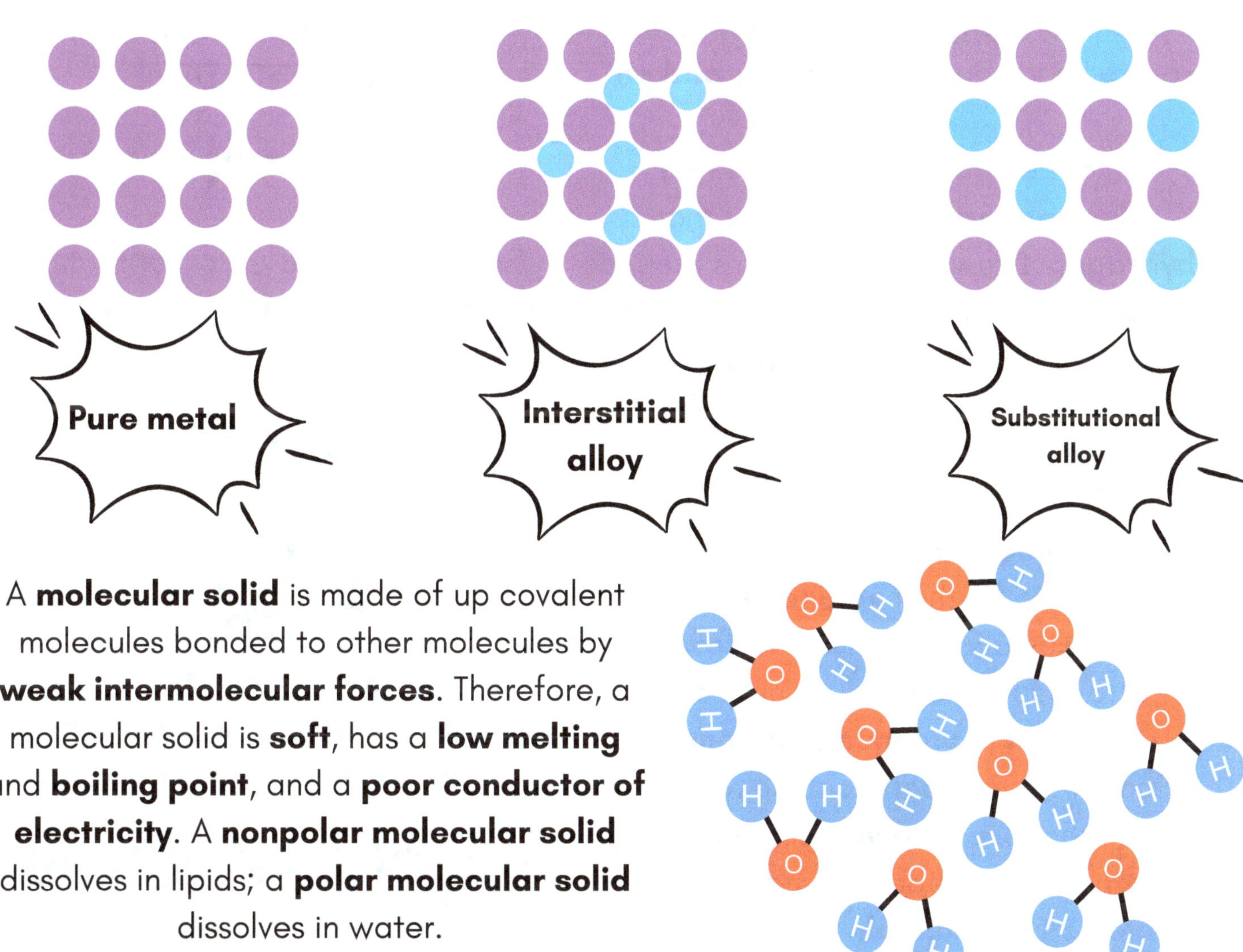

A **molecular solid** is made of up covalent molecules bonded to other molecules by **weak intermolecular forces**. Therefore, a molecular solid is **soft**, has a **low melting** and **boiling point**, and a **poor conductor of electricity**. A **nonpolar molecular solid** dissolves in lipids; a **polar molecular solid** dissolves in water.

A **Lewis diagram** is a method of drawing **molecules** and **compounds** with their valence electrons: **Here are some examples!**

To make a Lewis diagram, follow the steps below:

1. **Sum** up the number of valence electrons. If the overall compound is **positive, subtract** an electron; if the overall compound is **negative, add** an electron.

2. Find the **central atom**, the **least electronegative atom** other than **hydrogen**. Draw the other atoms around and connect them to the central atom with **single bonds. Subtract** the electrons consumed with the bonds from the number of electrons from **step 1**. Each bond is worth **two electrons**.

3. Add **lone pairs** to the **non-central atoms** to **fill** their **valence shells**.

4. If necessary, **assign** any **leftover electrons** to the central atom to achieve an **octet (8 valence electrons)**. Sometimes, the octet can be reached with **more** or **fewer** electrons.

5. If the central atom doesn't have an octet, create **multiple** bonds until ALL atoms have octets.

Sum of valence electrons: 4 + 5 + 1 = 10
No central atom!

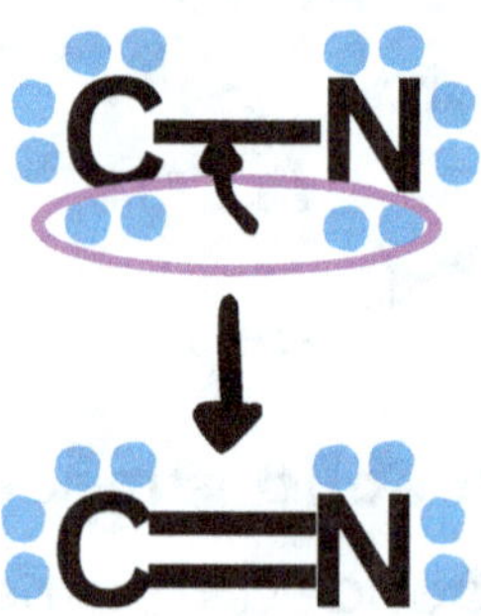

Sum of above electrons: 14, which is four more electrons more than we have.

Sum of above electrons: 12, which is two more electrons more than we have.

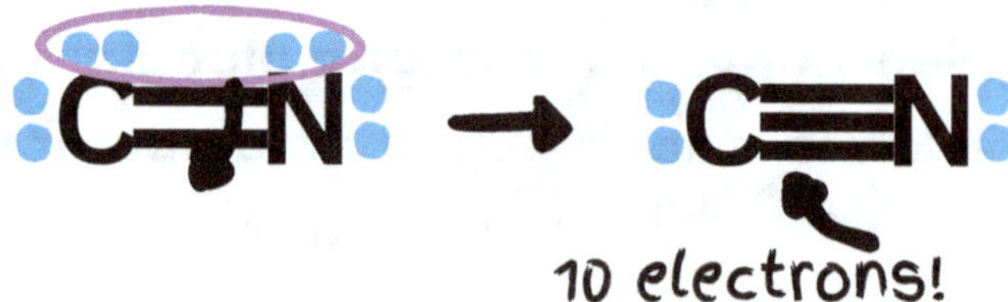

EXCEPTIONS!

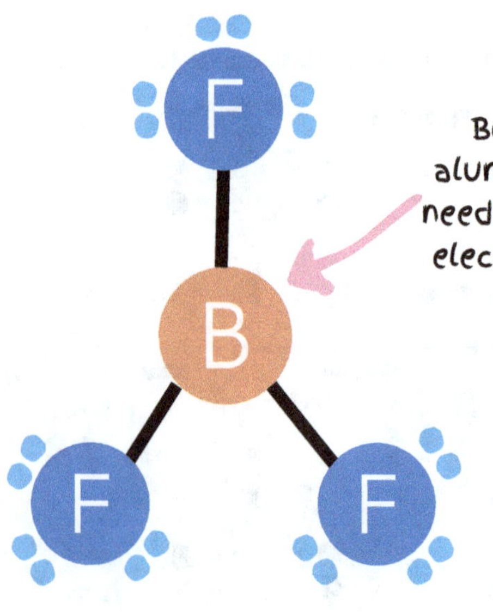

Boron and aluminum only need six valence electrons to be stable!

Elements like chlorine, xenon, sulfur, phosphorus, etc can expand their octets with their empty orbitals!

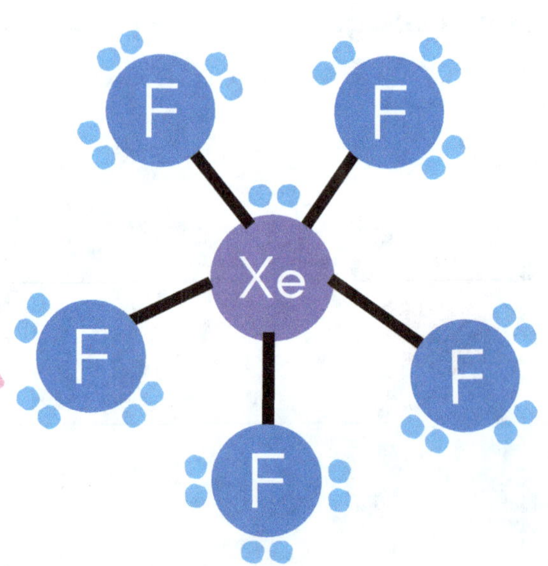

Molecules can exhibit **resonance**, which is when there are **multiple** possible Lewis diagrams to display a molecule. For example, the **carbonate ion**.

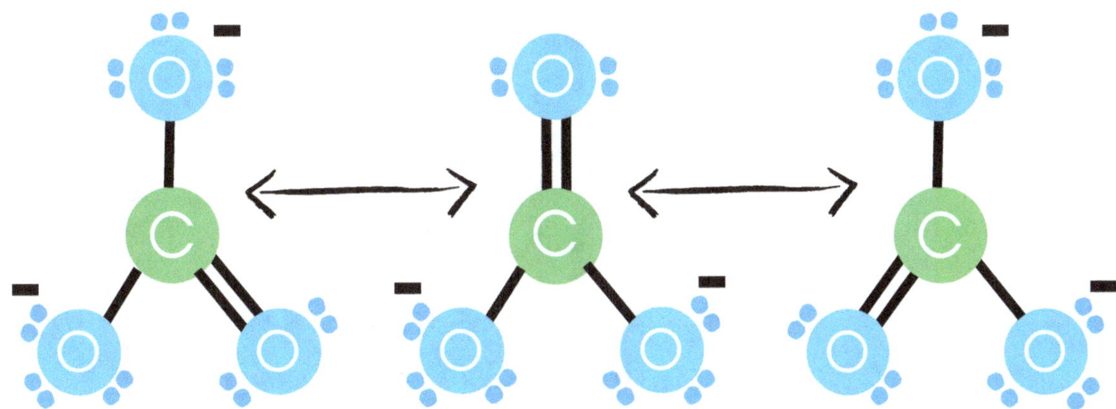

The **strength** and **length** of the **bonds** are **identical** because the double bond can be "**shared**" across all three bonding regions in the molecule. Add the **bond orders (single: 1, double: 2, triple: 3)** in one molecule **(in this case, 1 + 1 + 2 = 4)**, and **divide** by the number of bonding regions **(3)**.

4/3 = 1.33

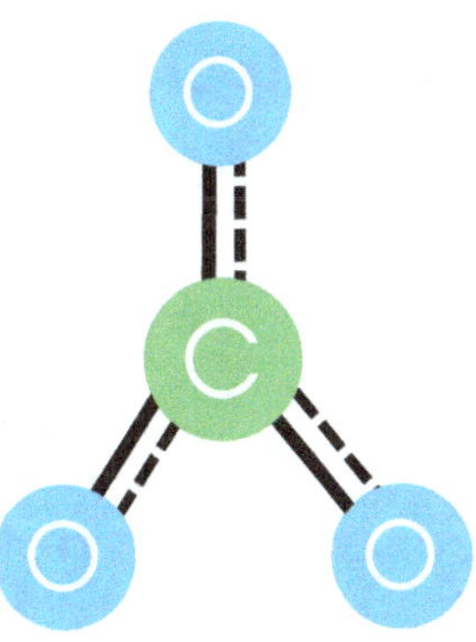

To find the **formal charge** of an atom in a compound, you can **subtract** the electrons that are within a bond **(lone pair electrons = 2, bonded electrons = 1)** from the **expected** number of electrons that are available to bond, which is found on the periodic table. There are two important concepts, however: the formal charges of all atoms must be as **close to zero** as possible, and the **sum** of all formal charges must be equal to the **charge** of the compound overall. If you look at the diagrams at the top of this page, the **formal charge** of the central atom is **4 - 4 = 0**, which is ideal. If there are **negative formal charges**, they would belong to the more **electronegative** atom.

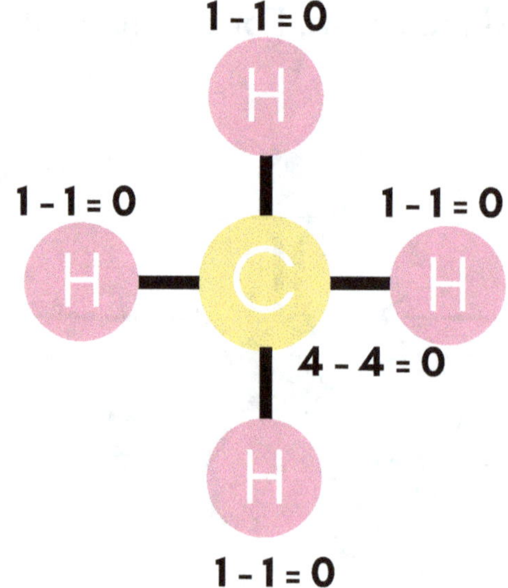

Intermolecular forces (IMFs) are the forces between molecules in a **covalently-bonded substance**. They need to be broken for substances to **change phases (physical change)**. When **ionic substances** change phase, the bonds between each **ion** are broken (**_intra_molecular forces**) whereas when **covalent substances** change phase, the bonds between the molecules are broken apart (**_inter_molecular forces**). The IMFs are **London-dispersion, dipole-dipole,** and **hydrogen bonding**.

Dipole-dipole forces occur between **polar molecules** when positive areas of molecules are attracted to negative areas of other molecules. With **greater polarity** comes **greater dipole-dipole attraction** and **higher melting** and **boiling points**. Substances with these forces **melt** and **boil at low temperatures**. An **ion-dipole force** is when a **negatively-charged atom** of one dipole molecule is attracted to a **small positive ion**.

Dipole-Dipole

A **hydrogen bond** is a type of dipole-dipole force in which the **positively-charged hydrogen** is attracted to the **negatively-charged end** of the **fluorine, oxygen,** or **nitrogen (FON)** of another molecule. These bonds are **stronger** than a normal dipole-dipole force because the **positively-charged nucleus** of the hydrogen is unshielded when it loses an electron.

London dispersion forces occur when electrons of a molecule randomly move towards one side, creating **sudden but temporary polarity**. As a molecule becomes **larger**, it will have **greater London dispersion forces**. Substances with London dispersion forces will melt and boil at **extremely low temperatures.**

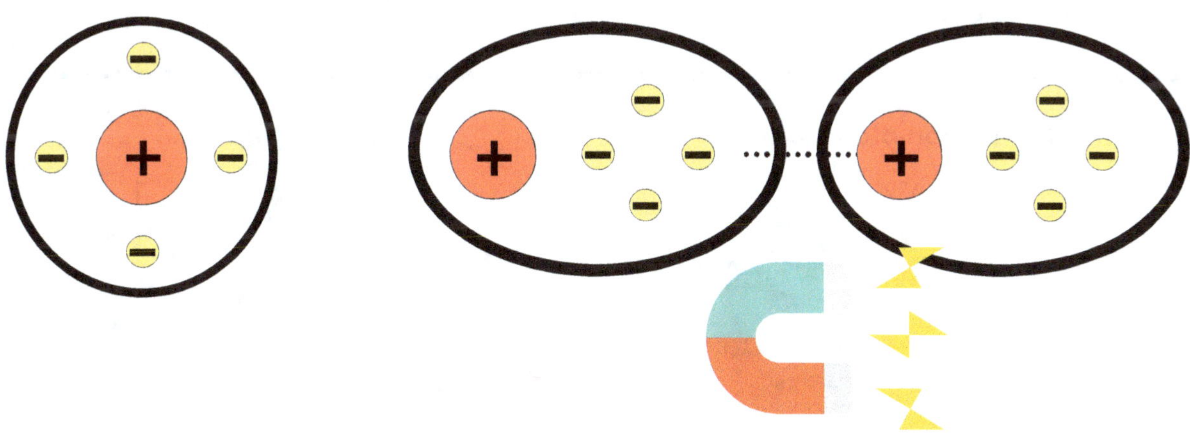

04 VSEPR SHAPES

In chemistry, electrons in the bonds of molecules will try to stay **as far apart as possible**. Molecules will take the shape that will allow their **bonds (and thus electrons)** to be as far away from each other as possible. When we predict the shapes of molecules using this idea, we are using the **valence shell electron-pair repulsion (VSEPR)** model.

Double and **triple** bonds are considered the same as **single** bonds in terms of predicting overall shape, but multiple bonds have *slightly more repulsive strength* and will occupy a *little more space* than single bonds. **Lone electron pairs** have more repulsive strength than bonding pairs, so molecules with lone pairs will have *slightly reduced bond angles* between atoms.

There are **electron-pair geometries** and **molecular geometries.** The **electron-pair geometry** is the shape of the molecule in terms of how many **electron clouds** (lone pairs, bonds) the central atom has. The **molecular geometry** takes into account different orientations of lone pairs and regular bonds. There is also **hybridization** of orbitals, which we will discuss later. The following shapes are the VSEPR shapes:

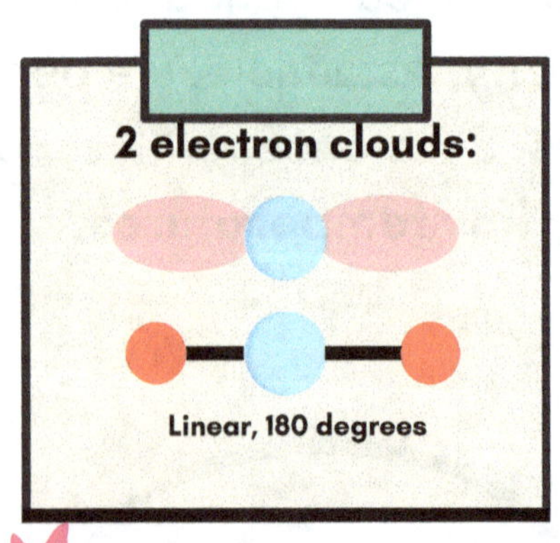

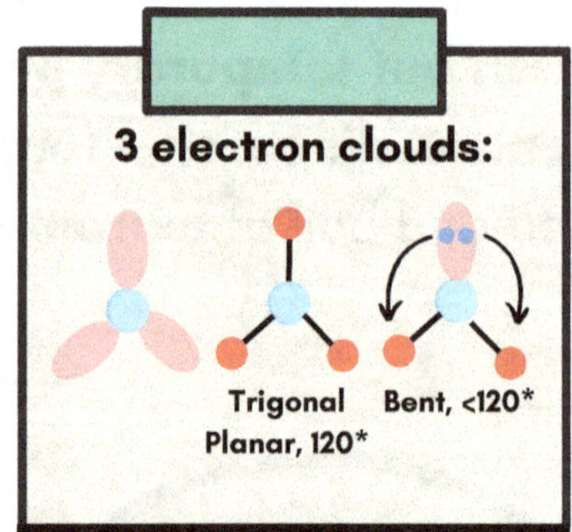

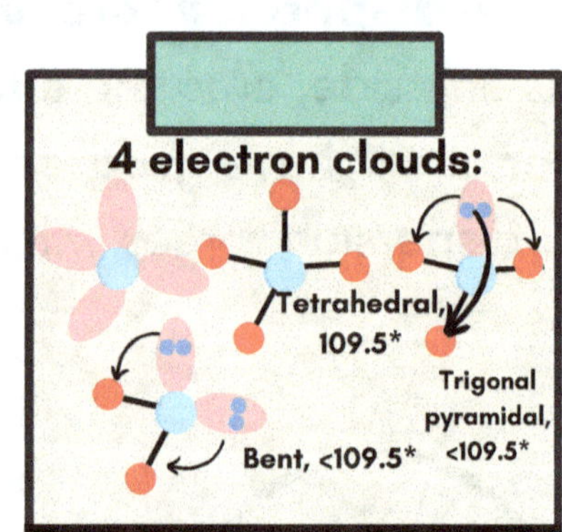

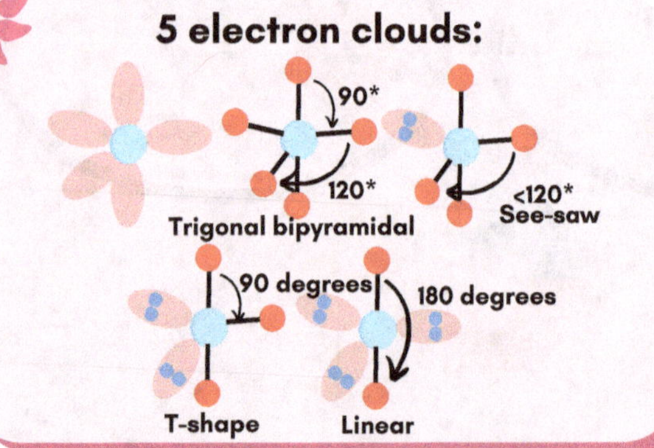

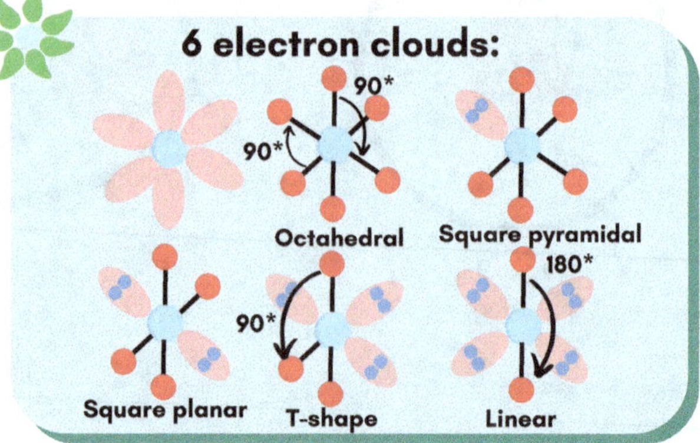

In chemistry, **bond hybridization** occurs when different orbitals mix to form new unique orbitals. The **Steric number** tells us the number of hybrid orbitals and is calculated by adding the **number of lone pairs** of the central atom (**two electrons = one pair**) to the number of bonds (**total number of electron-dense regions**).

sp **hybridized:** one **s** and one **p** character; **linear**

sp² **hybridized**: one **s** and two **p** characters; **trigonal planar**

sp³ **hybridized**: one **s** and three **p** characters; **tetrahedral**

***There is also *sp³d*, but we won't get into it.

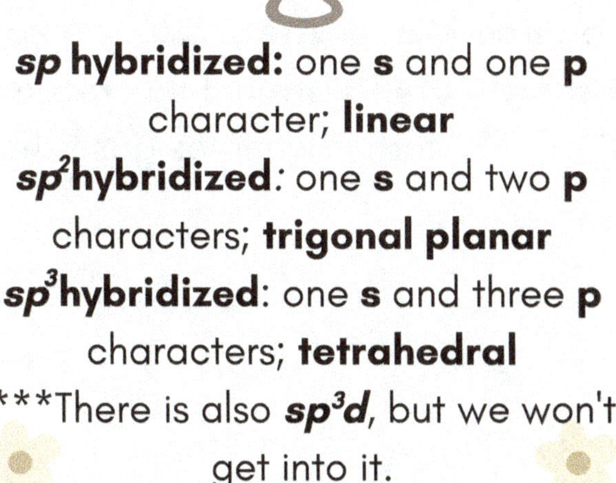

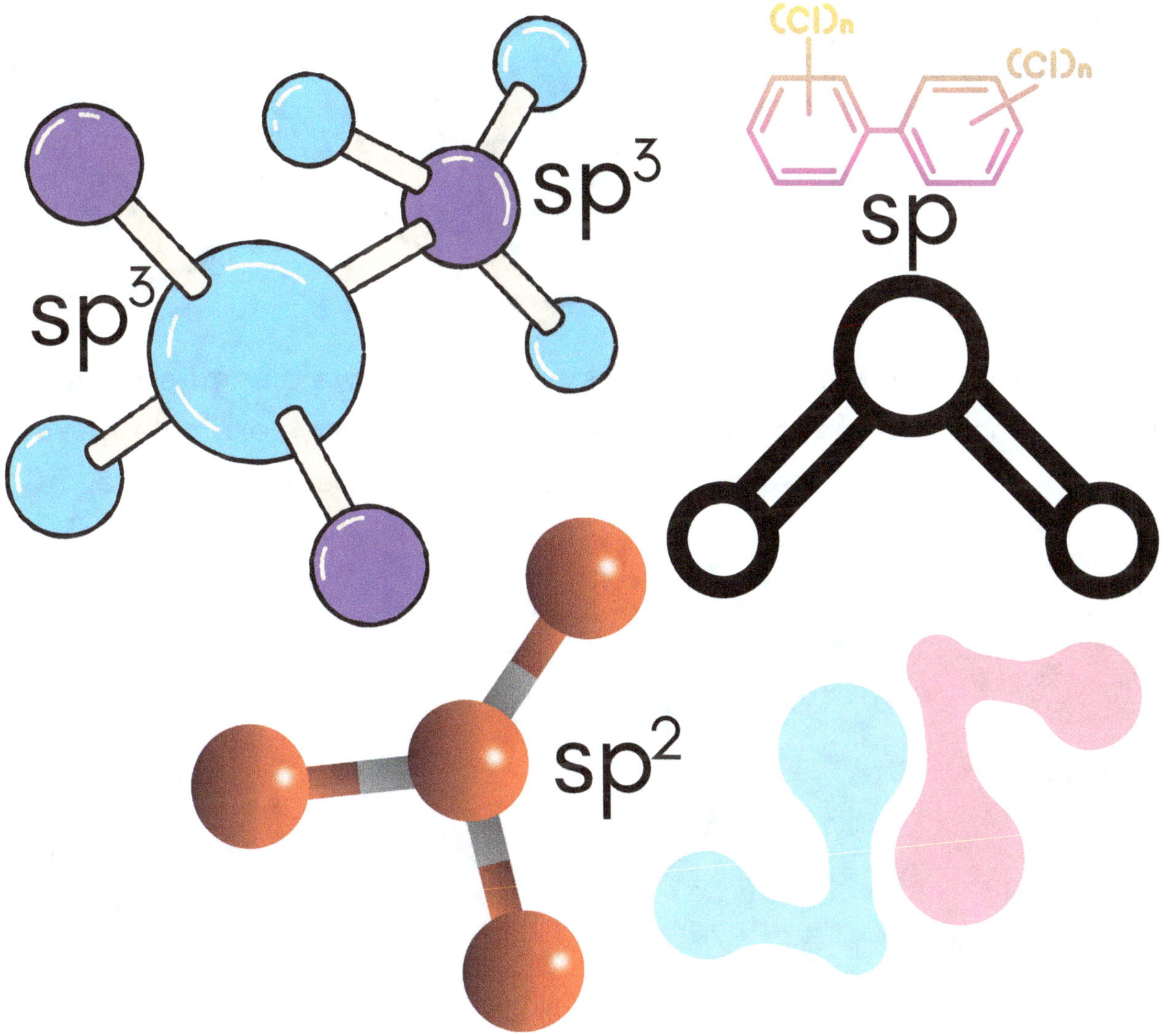

sp³

sp³

sp²

05 PHASES & LIGHT

The **phases of matter**, including **solid, liquid,** and **gas**, are determined by the **strength of intermolecular forces (IMFs)**. Solids have strong intermolecular forces whereas gases have weak intermolecular forces.

Amorphous solids are flexible substances that have a **semi-solid** structure. They are made up of **tangled polymers**. **Rubber** is one example. **Crystalline** solids form when the polymers align in a specific structure.

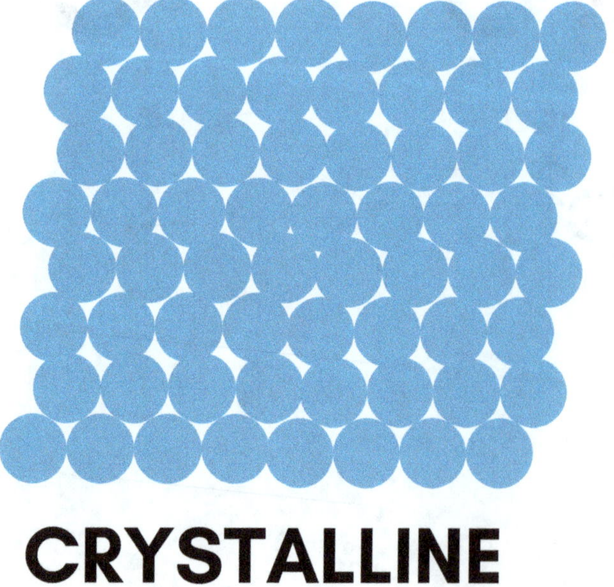

CRYSTALLINE

AMORPHOUS

Let's talk about **vapor pressure**. **Vapor pressure** results when molecules in motion within a **liquid** hit the surface of the liquid with enough **energy** and thus move into the **gas phase**. The *greater* the **temperature** of a liquid, the *faster* the molecules move and the *more likely* they are to break free of the other molecules (**higher vaporization/vapor pressure**). If two liquids are at the **same temperature**, then the **strength of the IMFs** within those two liquids determines which liquid will have a higher vapor pressure: the *stronger the IMFs*, the *lower* the vapor pressure.

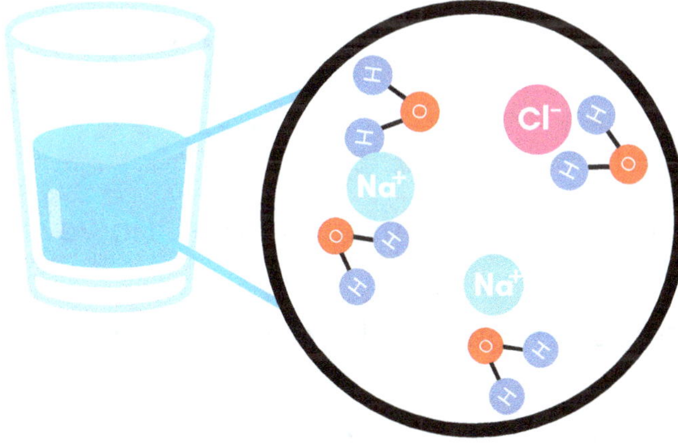

Ionic, or **polar**, **solutes** (substance that dissolves) dissolve in **polar solvents** (substance that does the dissolving). As a result, **salt (NaCl, ionic)** dissolves in **water**, which is polar. **Nonpolar solutes** dissolve in **nonpolar solvents**. As a result, **salt (NaCl, ionic)** does **not** dissolve in **oil.** When an ionic substance dissolves, it breaks up into ions, which are called **electrolytes** because they can conduct electricity. This is called **dissociation**.

Chromatography is the separation of a mixture depending on **polarity**.

PAPER

COLUMN

Speaking of **vaporization**, let's talk about **gases**. A **gas** is a state of matter that can be classified as either **ideal** or **real**. An **ideal gas** has many properties that are worth knowing:

When ideal gas particles move around, they can hit each other. However, the particles don't interact or attract each other, which results in elastic collisions. Additionally, the collective volume of the particles is so tiny that it is assumed, in simple chemistry, that the volume doesn't matter.

The kinetic energy depends on mass and temperature.

$$KE = 1/2\ mv^2$$

m = mass, v = velocity

The **equation** used to represent **ideal gas** is known as, obviously, the **ideal-gas equation:**

PV = nRT

P = pressure, **V** = volume, **n** = number of moles of gas, **R** = ideal-gas constant, **T** = temperature (in **Kelvin**, which is determined by adding **273** to whatever **degrees-Celsius**

R = 0.0821 L-atm/K-mol, or **8.314 J/K-mol**

You can also manipulate the ideal-gas equation to obtain the **Combined Gas Law: $P_1V_1 = P_2V_2$** when the number of moles of gas is held constant.

<u>Dalton's Law</u> states that the **pressure** of a mixture of gases can be discovered by adding up the individual pressures of all the gases in the mixture.

$$P_{total} = P_A + P_B + P_C + ...$$

The **partial pressure** of a gas depends on how many moles of the gas there are in the mixture.

$$P_A = (P_{total})(X_A)$$

X_A is the **mole fraction**: (number of moles of gas A) / (number of total moles)

There are **three laws** that are derived from the ideal-gas equation:

1. **<u>Gay-Lussac's Law</u>:** *As pressure increases, temperature increases & vice-versa.*
2. **<u>Boyle's Law</u>:** *As pressure increases, volume decreases & vice-versa.*
3. **<u>Charles's Law</u>:** *As temperature increases, volume increases & vice-versa.*
4. **<u>Avogadro's Law</u>:** *As the number of moles of a gas increases, the volume increases & vice-versa.*

A method of depicting the relationship between **molecular mass** and **speed** of gas molecules is the **Maxwell-Boltzmann diagram**. If multiple gases are in a sample at a **given temperature**, all of the gases will have the **same average kinetic energy**. Regardless of the type of gas, the kinetic energy is the same because it depends on temperature. However, the **velocity (speed distribution)** of the gases are not always the same at given temperatures. There are **two** types of diagrams: **one type of gas at multiple temperatures** or **multiple types of gas at the same temperature**.

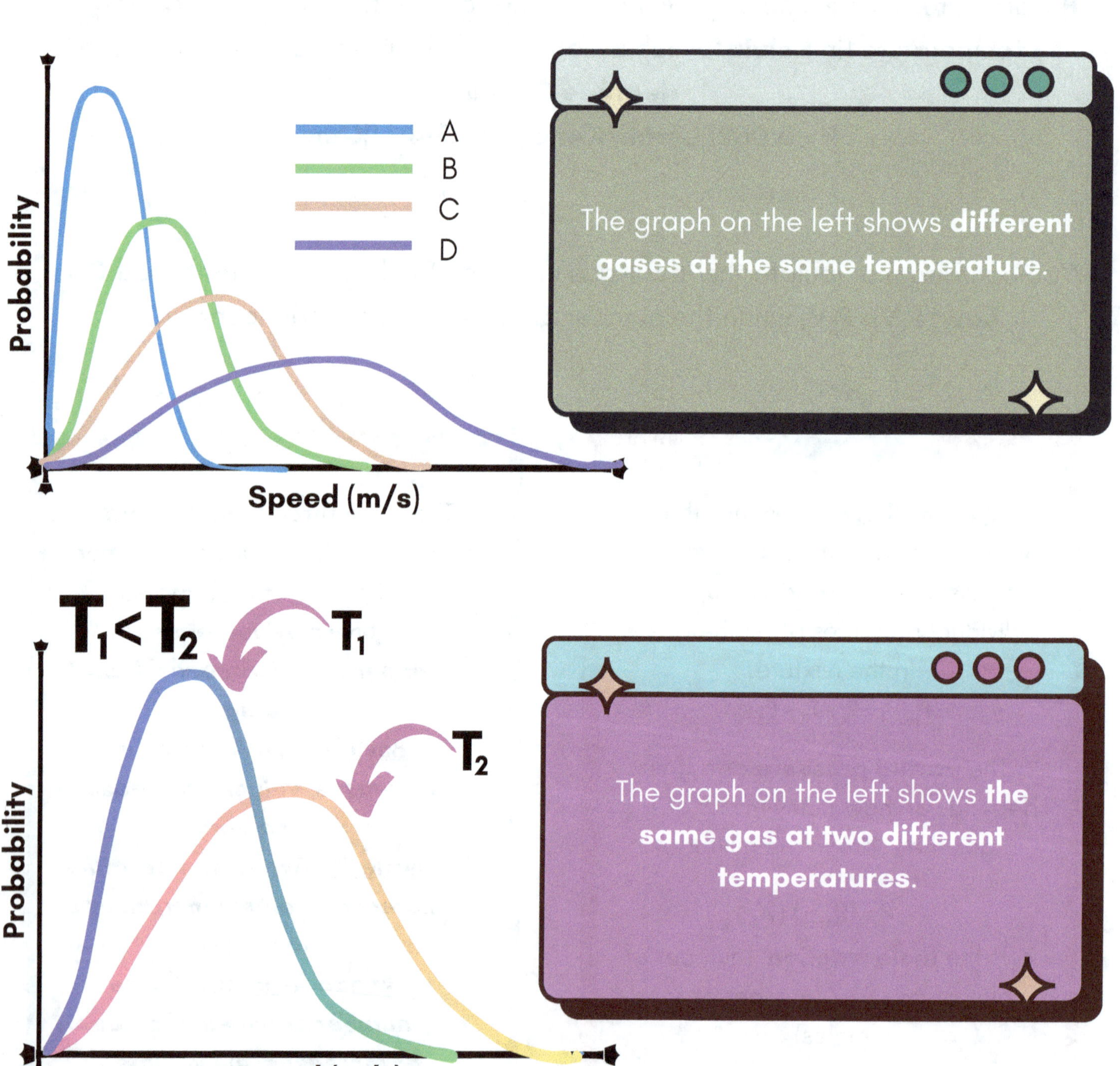

When a balloon is filled with gas, the balloon does not stay perfectly inflated forever: over time, the balloon starts to **deflate** because the **gas molecules in the balloon escape** through **tiny holes on the balloon**. The rate at which gas molecules escape through the tiny holes is **effusion.** This depends on how fast the molecules move and how much the molecules weigh: the **smaller** the molecule, the **faster** it will diffuse.

There is an equation to know and understand so that you can calculate the **ratio** between the **effusion rates** of **two gases.**

$$\frac{\text{Rate}_{\text{gas 1}}}{\text{Rate}_{\text{gas 2}}} = \sqrt{\frac{\text{Molar Mass}_{\text{gas 1}}}{\text{Molar Mass}_{\text{gas 2}}}}$$

Faster?

At a **high pressure** or a **low temperature**, gas molecules tend to be closer together and thus interact with each other, which goes against ideal behavior. For gas to be completely ideal, the gas molecules must never interact with each other.

A **mixture** can be a **solid, liquid**, or **gas**. A **homogenous mixture** looks like **one substance** because the substances are equally distributed, and **solutions** are great examples of homogenous mixtures. In a solution, there is a **solute** that dissolves in a **solvent**. Some examples of homogenous mixtures are air, saltwater, etc. A **heterogeneous mixture** is **visibly different**. It is made up of different substances and can be separated quite easily. Some examples of heterogeneous mixtures are **sand**, **chicken noodle soup**, and **oil** in **water**.

We talked about **solutions**, but they are often referred to as **aqueous solutions** for specifics.

Molarity is used to measure the amount of solute in a solution. It is represented by **moles of solute** over **liters of solvent (mol/L = _M_)**. Using molarity, we can solve **dilution problems**.

Dilution is a process of **lowering the molarity** of a solute in a solvent. We do this by adding more solvent to the solution. We use the equation $M_1V_1 = M_2V_2$ to solve such questions.

How many milliliters of water should be added to dilute 500.0 mL of a 0.400 M NaCl solution to a 0.200 M NaCl solution?

$$M_1 V_1 = M_2 V_2$$

$$(0.400\ M)(500.0\ mL) = (0.200\ M)(V_2)$$

$$V_2 = 1000.0\ mL$$

We need to add **500.0 mL** of water to dilute the solution.

A **photon** makes up **light**. Electrons **absorb** photons and gain energy. After a while, these electrons **release** the photon as they lose the energy they gain. **Spectroscopy** deals with the relationship between light and matter.

There are a couple of equations that you need to know:

$$E = h\nu$$

E = energy (**J** = **Joules**)
h = **Planck's constant** (6.626×10^{-34} J–s)
ν = frequency (1/s)

$$c = \lambda \nu$$

c = speed of light (2.998×10^{8} m/s)
λ = wavelength (m)
ν = frequency (1/s)

The first ionization energy of **lithium** is **519 kJ/mol**. What is the longest wavelength of light that can ionize an atom of lithium in gaseous form?

λ We are looking for wavelength!

519 kJ/mol = 519,000 J/mol

519,000 J/mol × 1 mol / 6.022×10^{23}

519,000 J / 6.022×10^{23}

= 8.62×10^{-19} J

$E = h\nu$, $\nu = E/h$
$c = \lambda\nu$, $\nu = c/\lambda$

$E/h = c/\lambda$

$\lambda = c/(E/h)$

$\lambda = ch/E$

$$\lambda = \frac{(2.998 \times 10^{8}\,\text{m/s})(6.626 \times 10^{-34}\,\text{J–s})}{8.62 \times 10^{-19}\,\text{J}}$$

$$= 2.30 \times 10^{-7}\,\text{m}$$

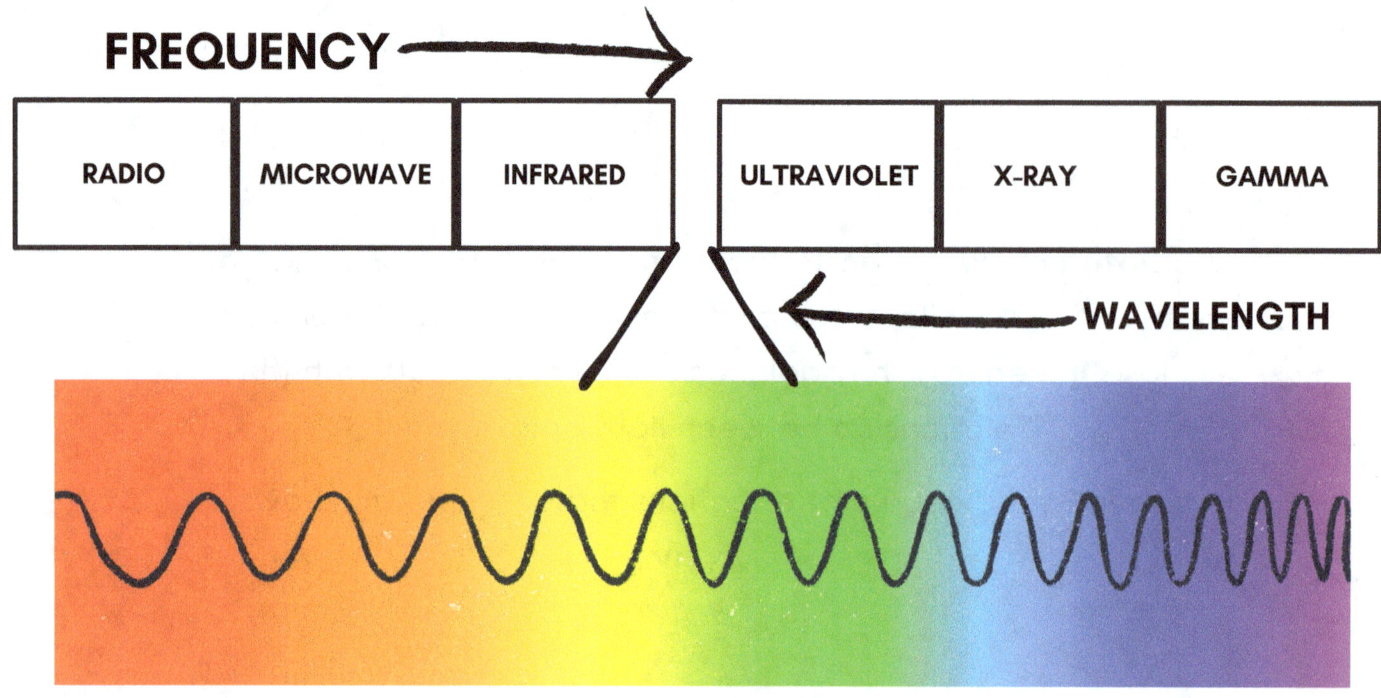

VISIBLE LIGHT

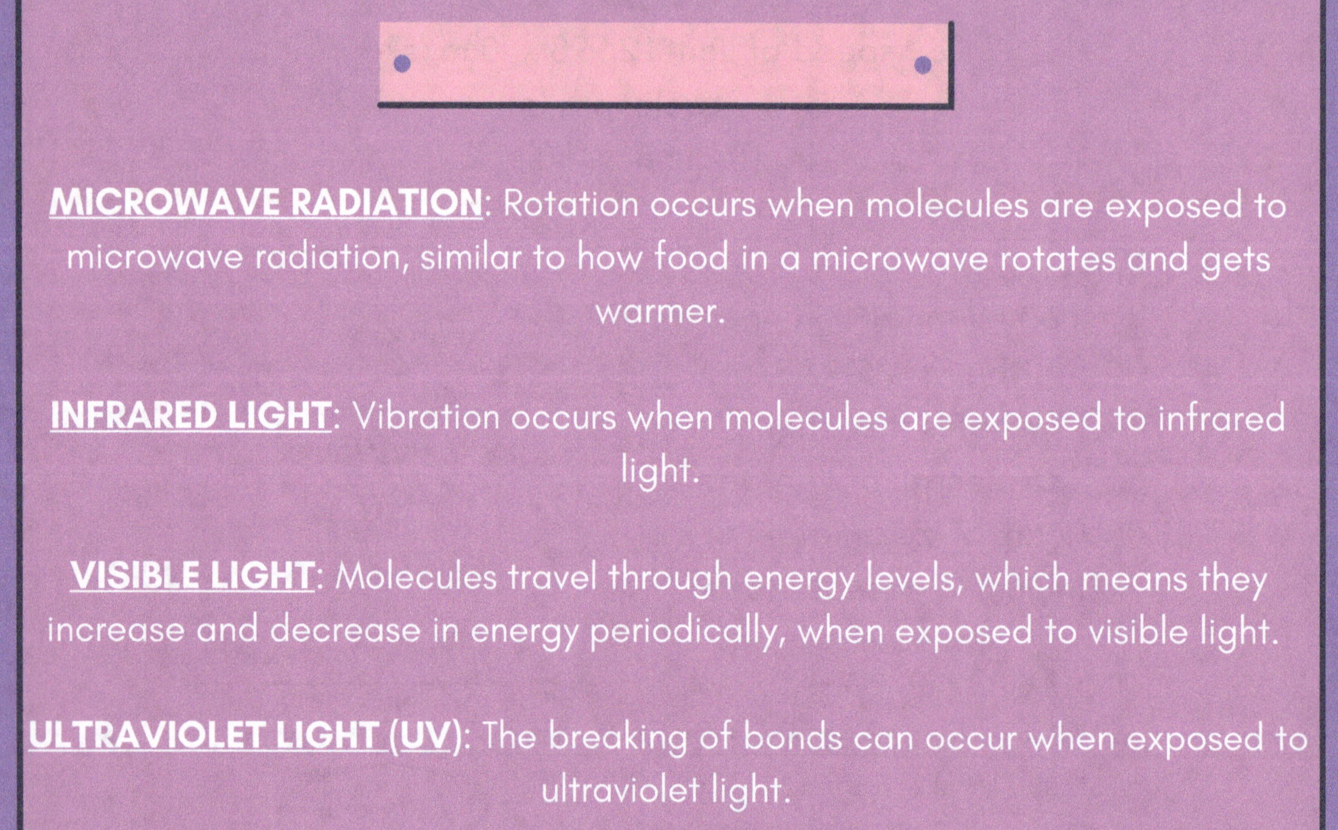

A **spectrophotometer** can be used to measure the **concentration** of a solution over time. This measures how much light at a given wavelength is **absorbed** by a solution. A good **indicator** that the amount of light that the solution absorbed changed is a **color change of the solution**. There is an equation that you need to know:

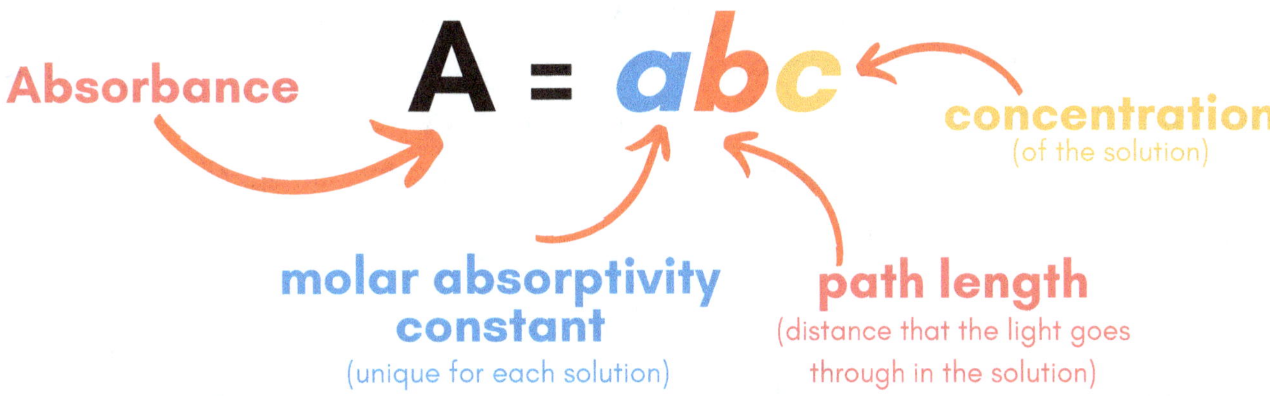

Beer-Lambert's law states that there is a relationship between **absorbance** and **concentration**, particularly in solutions with color.

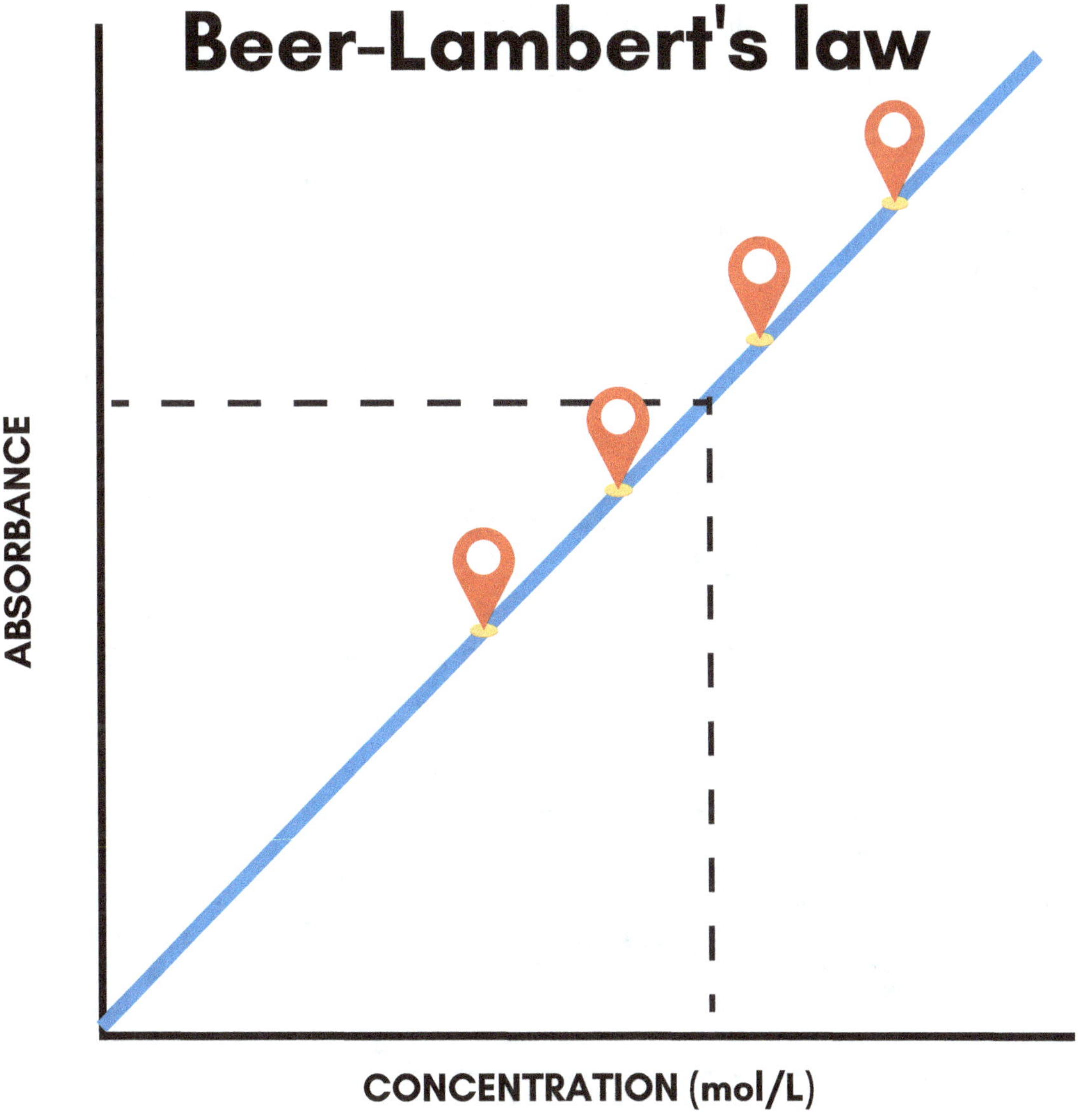

Beer-Lambert's law

06 REACTIONS

There are two types of **changes** in chemistry: **physical** and **chemical**. A **physical change** occurs when matter changes **form** and is **reversible**. Some examples of physical changes are cutting wood into smaller pieces, freezing/melting, boiling, etc. A **chemical change** occurs when **new products** are formed and is **irreversible**. Some examples of chemical changes are rotting, combustion, digestion, etc.

Synthesis Reaction: multiple *simple* compounds come together to make a *complex* compound.

Decomposition Reaction: a single compound produces two or more elements or simpler compounds.

Acid-Base Reaction: an acid reacts with a base to form water and a salt.

Redox Reaction: one compound loses electrons to become more positive and the other gains electrons to become more negative.

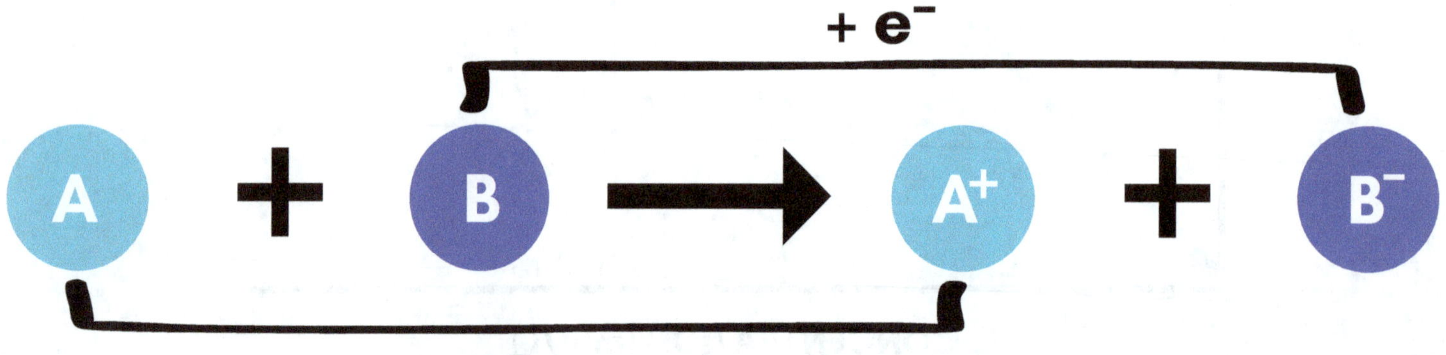

Combustion Reaction: a *covalent* substance with **carbon** and **hydrogen** is lit, which results in **carbon dioxide** and **water** by reacting with the oxygen in the atmosphere.

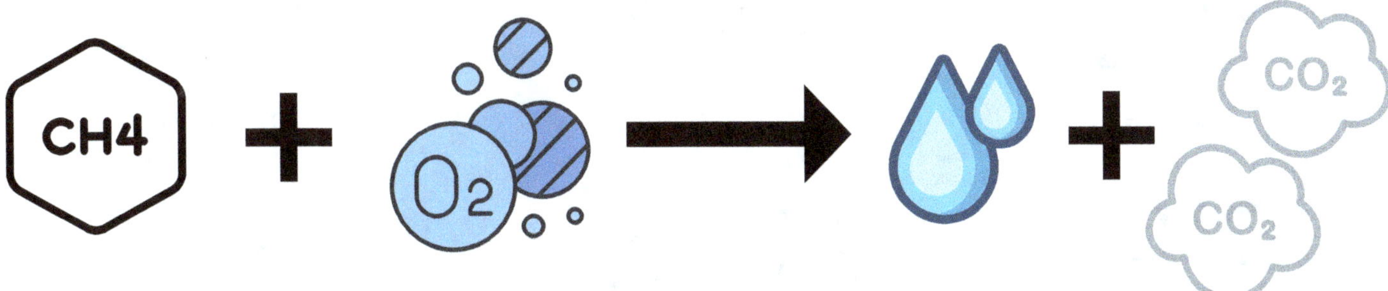

Precipitation Reaction: when two solutions are combined, a **solid** compound can form from the ions.

Single-Replacement Reaction: a lone element replaces another element in a compound.

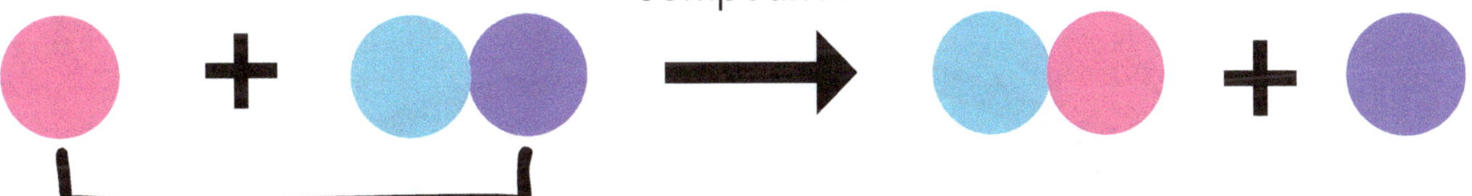

Double-Replacement Reaction: two compounds "trade" elements.

In a **single-replacement reaction**, the lone element must be **more reactive** than the element it replaces. In a **double-replacement reaction**, a **gas** or a **precipitate** always forms.

There are some **polyatomic ions** that are important to remember:

$$NO_3^-$$

nitrate

$$CO_3^{2-}$$

carbonate

$$NH_4^+$$

ammonium

$$ClO_3^-$$

chlorate

$$MnO_4^-$$

permanganate

$$SO_4^{2-}$$

sulfate

$$C_2H_3O_2^-$$

acetate

$$H_3O^+$$

hydronium

$$OH^-$$

hydroxide

Solubility refers to how much of a compound can dissolve in a solution. There are certain compounds that, when dissolved in solution, are **soluble** whereas others are **insoluble**. The following rules relate to soluble compounds:

07 KINETICS

In chemistry, all reactions occur at a certain **speed**, which depends on **energy**, **temperature**, and the **nature** of the reaction. The **rate** of the reaction depicts how fast or slow a reaction is. It is important to note that **rates** are **not constant** throughout the reaction because **reactants (substances at the beginning)** are used up. Reactions can be depicted using a **rate law**.

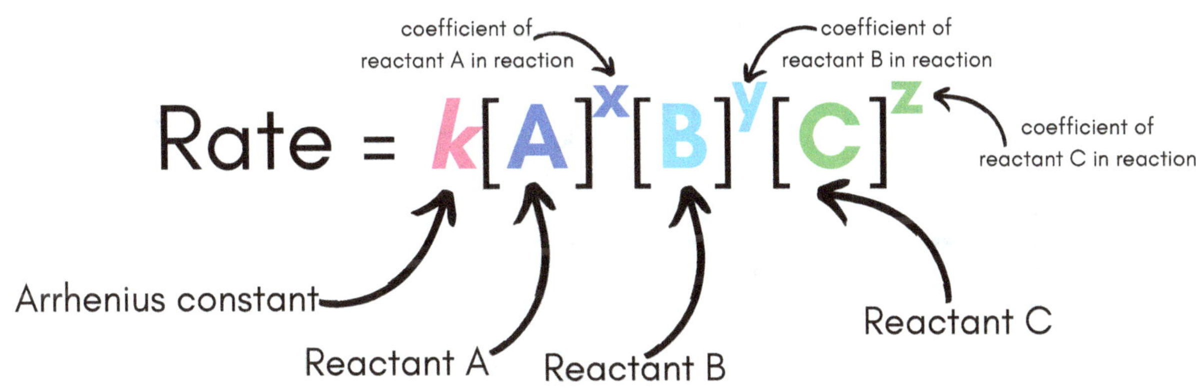

There are different **orders of reactions** as well as **orders of reactants**.

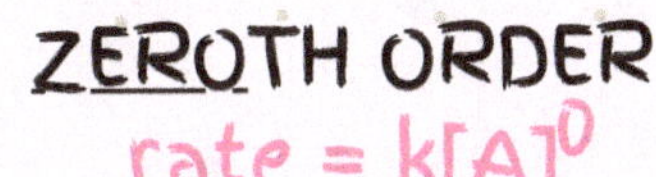

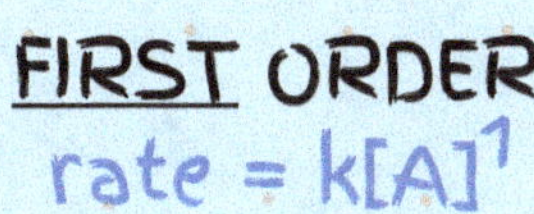

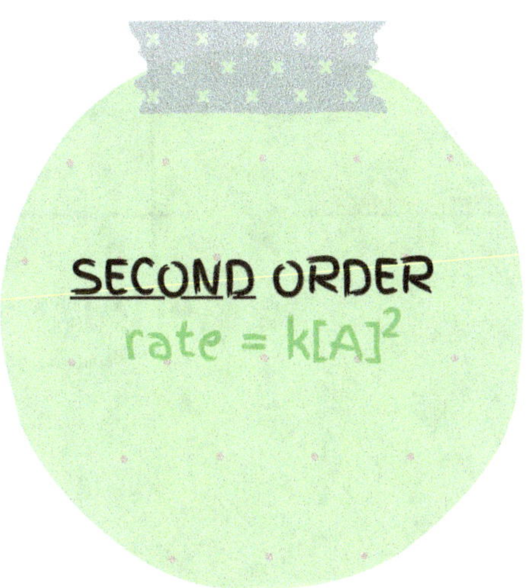

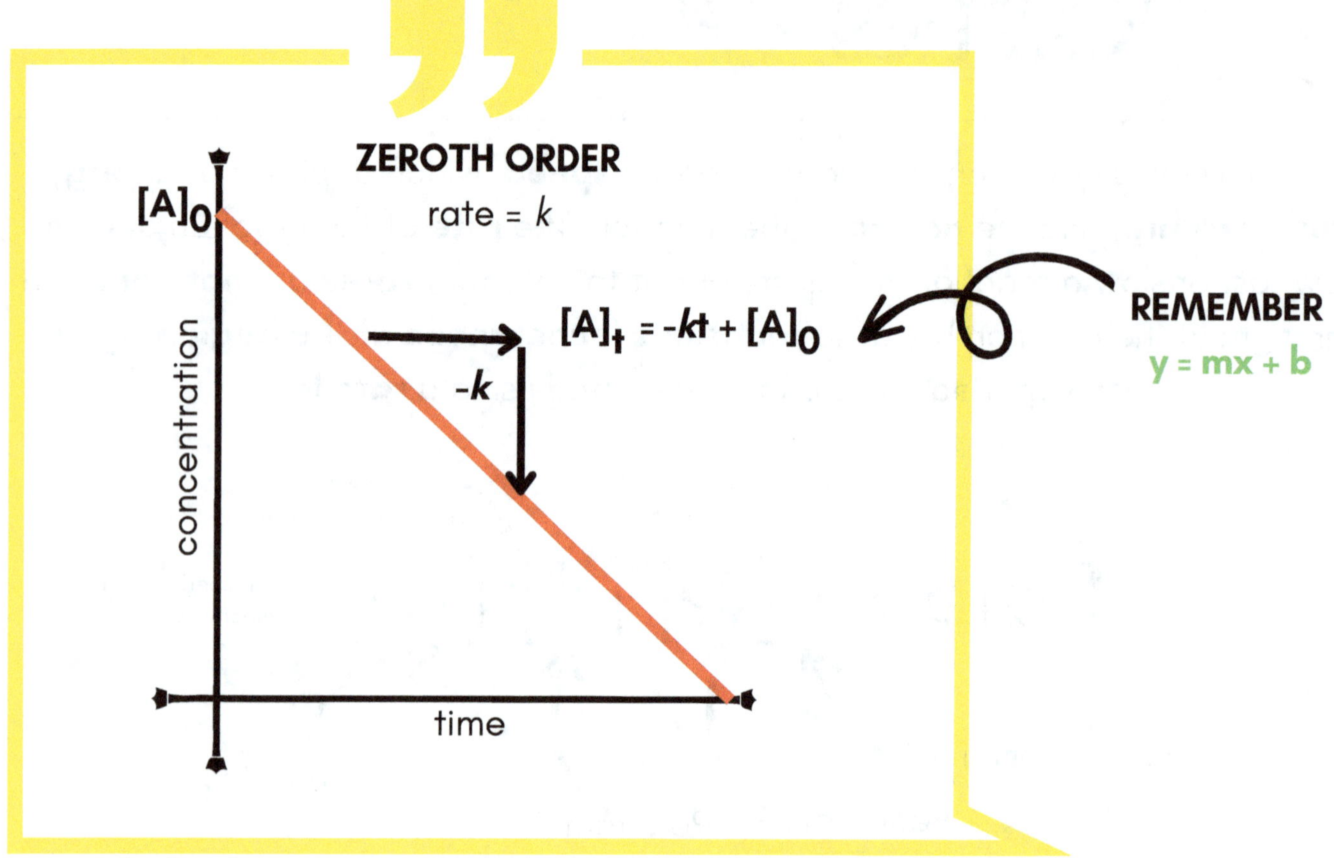

ZEROTH ORDER
rate = k
$[A]_0$
concentration
time
-k
$[A]_t = -kt + [A]_0$
REMEMBER
y = mx + b

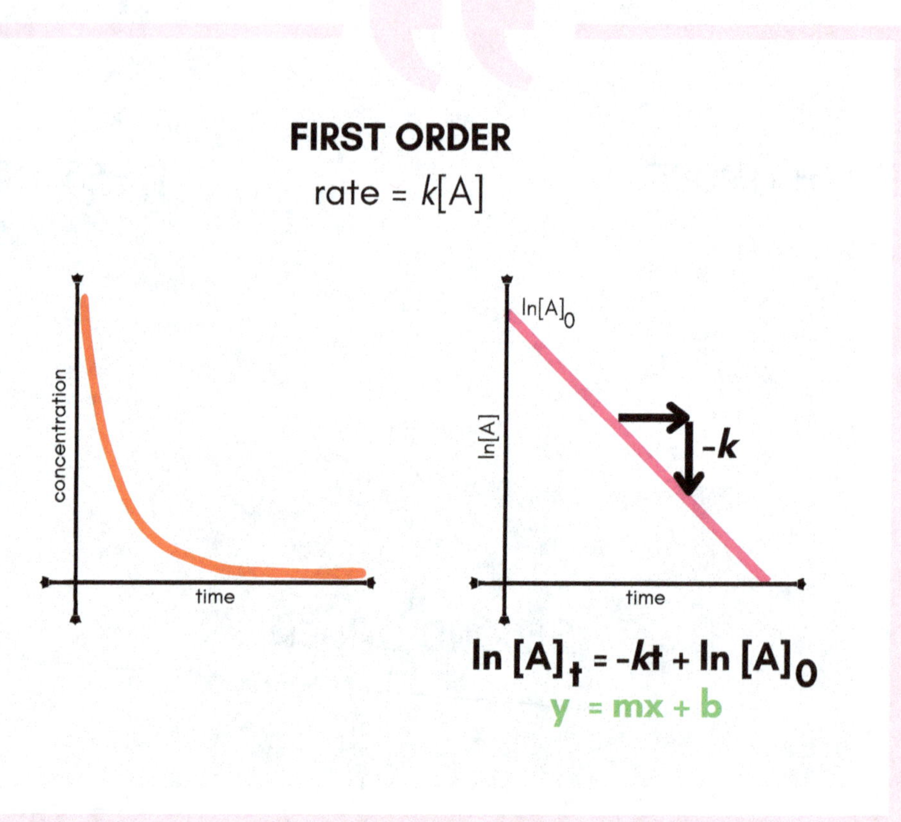

FIRST ORDER
rate = k[A]
concentration
time
$\ln[A]_0$
$\ln[A]$
time
-k
$\ln [A]_t = -kt + \ln [A]_0$
y = mx + b

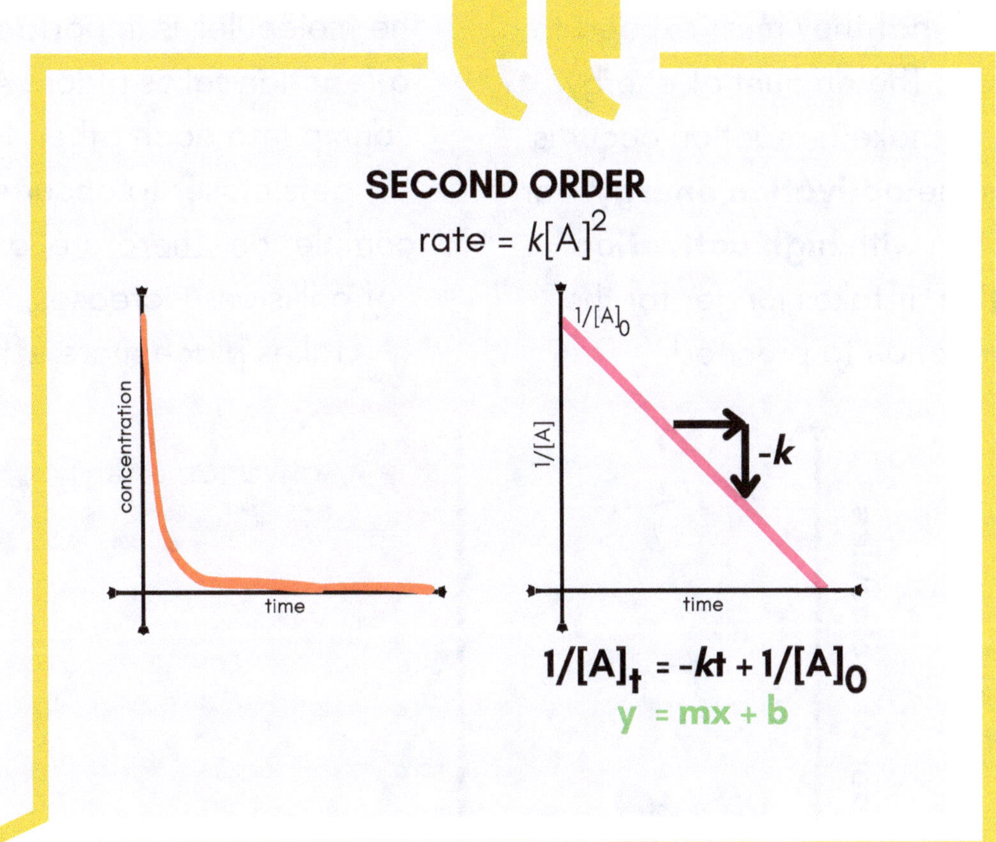

The **half-life** is how long it takes for **half** of a sample to decay. This is often used in terms of **radioactive samples**. In **first-order reactions**, the sample's half-life remains constant. An equation used to calculate half-life in such a scenario is **0.693/k**, and **k** is the **rate constant**.

All reactions have a certain **energy threshold** that they must exceed to continue. The amount of energy needed to make a reaction occur is known as the **activation energy**. For reactions with **high activation energies**, it takes longer for the reaction to proceed.

The **frequency of collisions** among the molecules is important to how fast a reaction takes place. As molecules bump into each other, the reaction gets closer to occurrence and completion. Therefore, as the number of collisions increases, the reaction takes place more effectively.

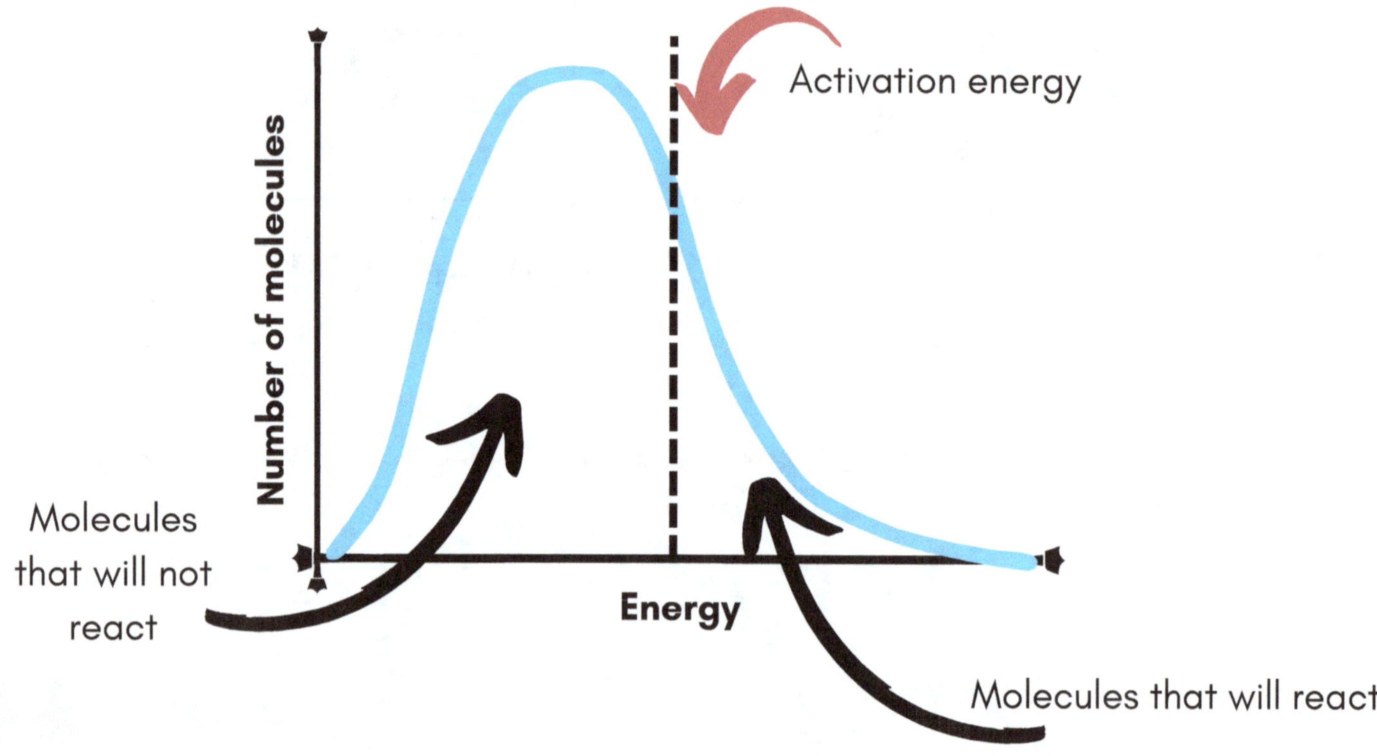

For a reaction to proceed faster, we can **increase the temperature, crush up the reacting substances, or stir the mixture!** We can also **add a catalyst**, which will offer an alternate way with a lower activation energy in which the reaction can take place.

08 THERMODYNAMICS

If you have **two** objects, one **hot** and one **cold**, the **hot** object transfers **heat** to the **cold** object until both objects have roughly the same amount of heat. **Temperature** is a numerical representation of how hot or cold an object is. Both concepts are important in **thermodynamics**.

In thermodynamics, there are three terms: **system**, **surrounding**, and the **universe**. The **system** is basically the chemical reaction itself, including the reactants and the products. The **surrounding** is anything besides the system. The **universe** includes both the **system** and **surrounding.**

We will discuss this more later, but **enthalpy** is a method of measuring how much energy is involved in a reaction. It is a numerical value, and its units are **Joules** or **kilojoules** because it represents energy. By looking at if the enthalpy of a reaction is **positive** or **negative**, we can analyze if energy was **absorbed** or **released**.

Since the actual enthalpy value of a reaction is difficult to calculate, we rely on the *change* in enthalpy.

There are two types of reactions in thermodynamics: **exothermic** and **endothermic**.

In an **exothermic reaction**, energy is **released**, which makes the enthalpy change **negative**, meaning the products have a **greater** enthalpy than the reactants.

When liquid water freezes, energy is released, which makes this reaction an **exothermic** reaction.

In an **endothermic reaction**, energy is **absorbed**, which makes the enthalpy change **positive**, meaning the products have a **lower** enthalpy than the reactants.

When ice melts, energy is absorbed, which makes this reaction an **endothermic** reaction.

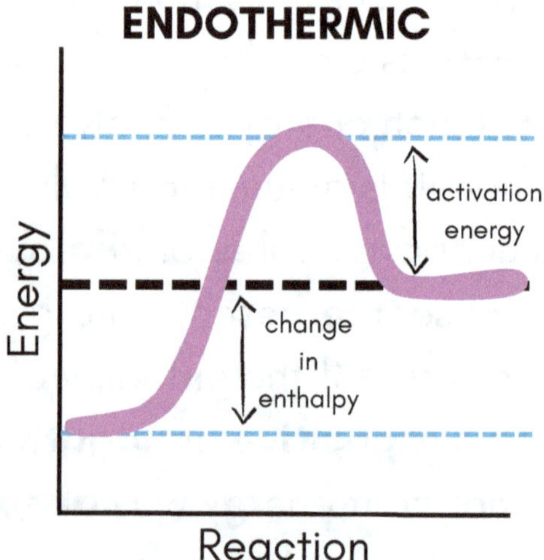

When a **catalyst** is added to the reaction, the reaction is able to occur **faster** because the catalyst introduces another path that the reaction can take. With this new pathway, the activation energy is lowered.

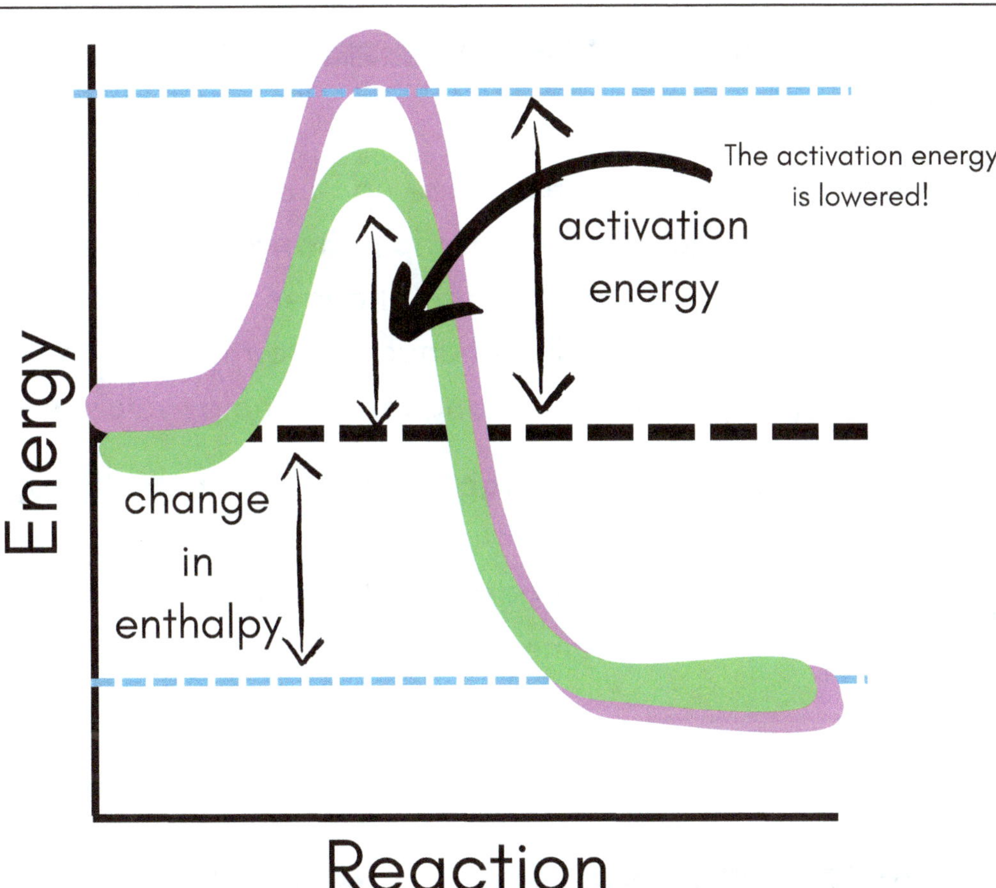

Contrary to the **enthalpy of reaction**, **bond enthalpy** is calculated by subtracting the **bond energy** of bonds **formed** from the **bond energy** of bonds **broken**.

If a complete reaction is made up of a combination of other reactions, the **overall enthalpy change** of the reaction is simply the **sum** of enthalpy changes of all of the other reactions. If you need to change the format of a constituent reaction, you need to change the enthalpy value as well.

If you **flip** a reaction, **negate** the sign of the enthalpy value.

If you **increase or decrease** the reaction by a factor, do the same for the enthalpy value.

This rule is known as **Hess's Law!**

$$CH_4 + 2\ O_2 \longrightarrow CO_2 + 2\ H_2O$$

	ΔH°
$C + 2\ H_2 \longrightarrow CH_4$	-74.8 kJ
$C + O_2 \longrightarrow CO_2$	-393.5 kJ
$H_2 + 1/2\ O_2 \longrightarrow H_2O$	-285.8 kJ

$CH_4 \longrightarrow C + 2\ H_2$	+74.8 kJ
$C + O_2 \longrightarrow CO_2$	-393.5 kJ
$2\ H_2 + O_2 \longrightarrow 2\ H_2O$	-571.2 kJ

$$CH_4 + 2\ O_2 \longrightarrow CO_2 + 2\ H_2O \qquad -890\ kJ$$

When a **state of matter** (solid, liquid, gas) turns into a different state of matter, a **phase change** is occurring. Specifically between **solids** and **liquids** as well as **liquids** and **gases**, there are specific terms to understand.

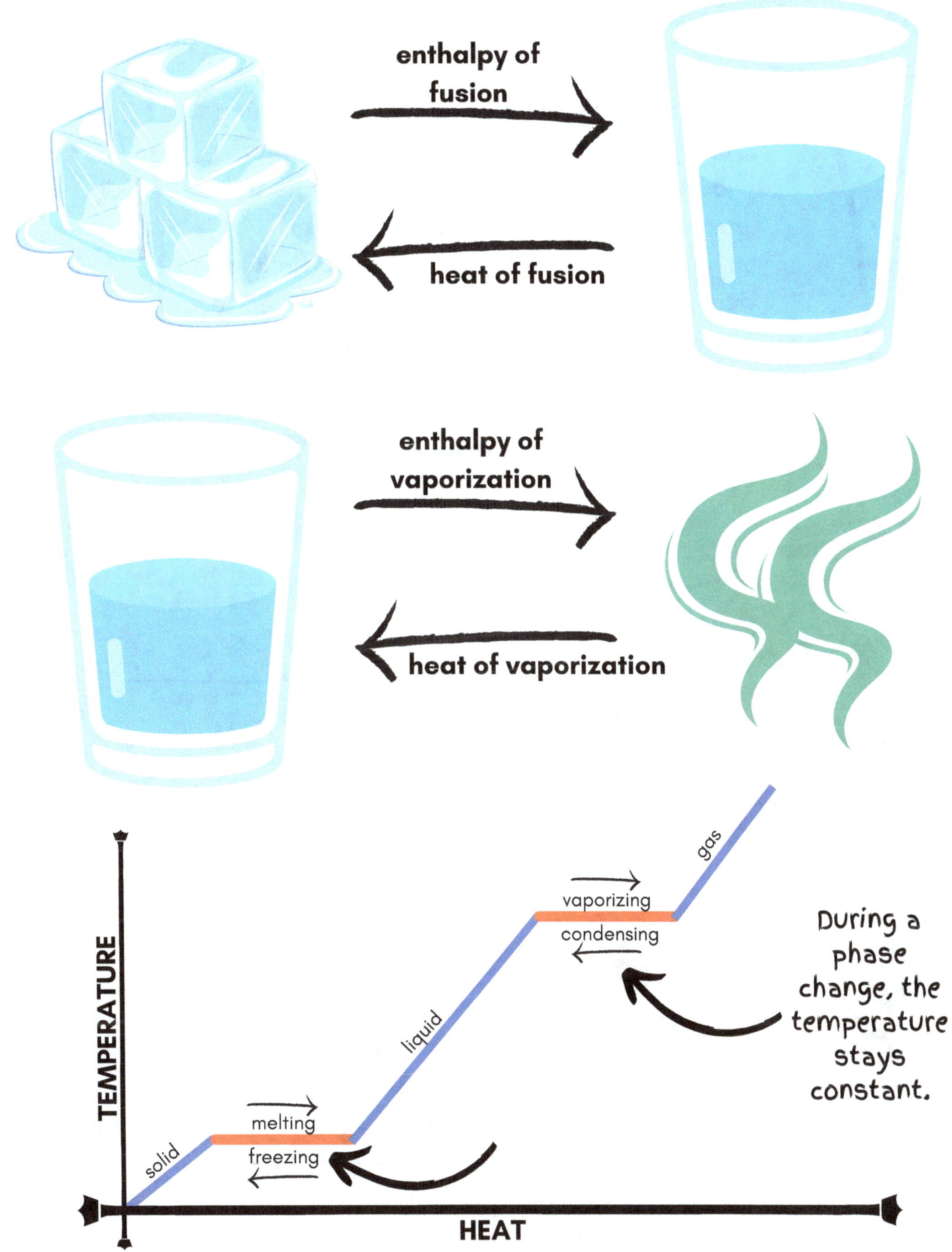

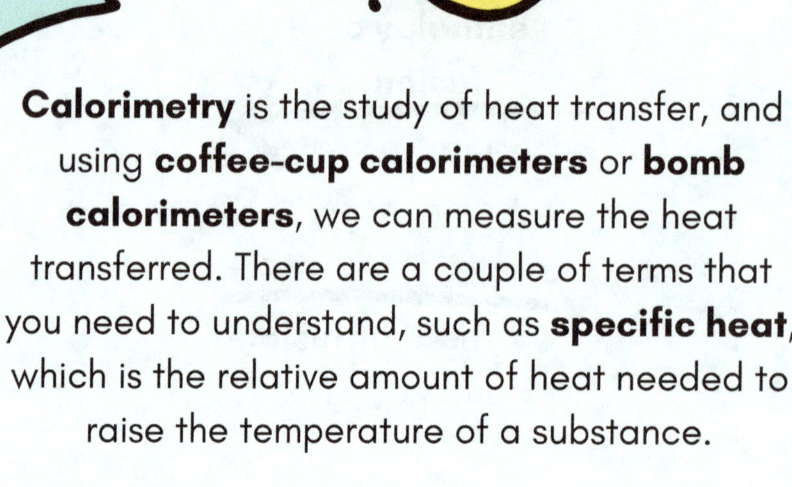

Calorimetry is the study of heat transfer, and using **coffee-cup calorimeters** or **bomb calorimeters**, we can measure the heat transferred. There are a couple of terms that you need to understand, such as **specific heat**, which is the relative amount of heat needed to raise the temperature of a substance.

$$q = mc\Delta T$$

q: heat (J or cal)

c: specific heat (J/g-celsius)

T: change of temperature

$$q_{phase\ change} = m\Delta H$$

q: heat (J or cal)

m: mass of substance

ΔH: heat of fusion, vaporization, etc

09 EQUILIBRIUM

In chemistry, at a certain point in time, a reaction reaches **equilibrium**, which is when the **forward reaction (reactants --> products)** occurs at the same rate as the **reverse reaction (products --> reactants)**. At this point, there is no noticeable change in the amount of reactants or products.

We use a special expression to portray equilibrium: the **equilibrium expression**.

$$K_{eq} = \frac{[C]^c [D]^d}{[A]^a [B]^b}$$

PRODUCTS

REACTANTS

coefficients in the reaction

concentrations or partial pressures of the substances in the reaction

The Haber process

$$3\ H_2\ (g) + N_2\ (g) \rightleftharpoons 2\ NH_3\ (g)$$

$$K_{eq} = \frac{[NH_3]^2}{[H_2]^3 [N_2]}$$

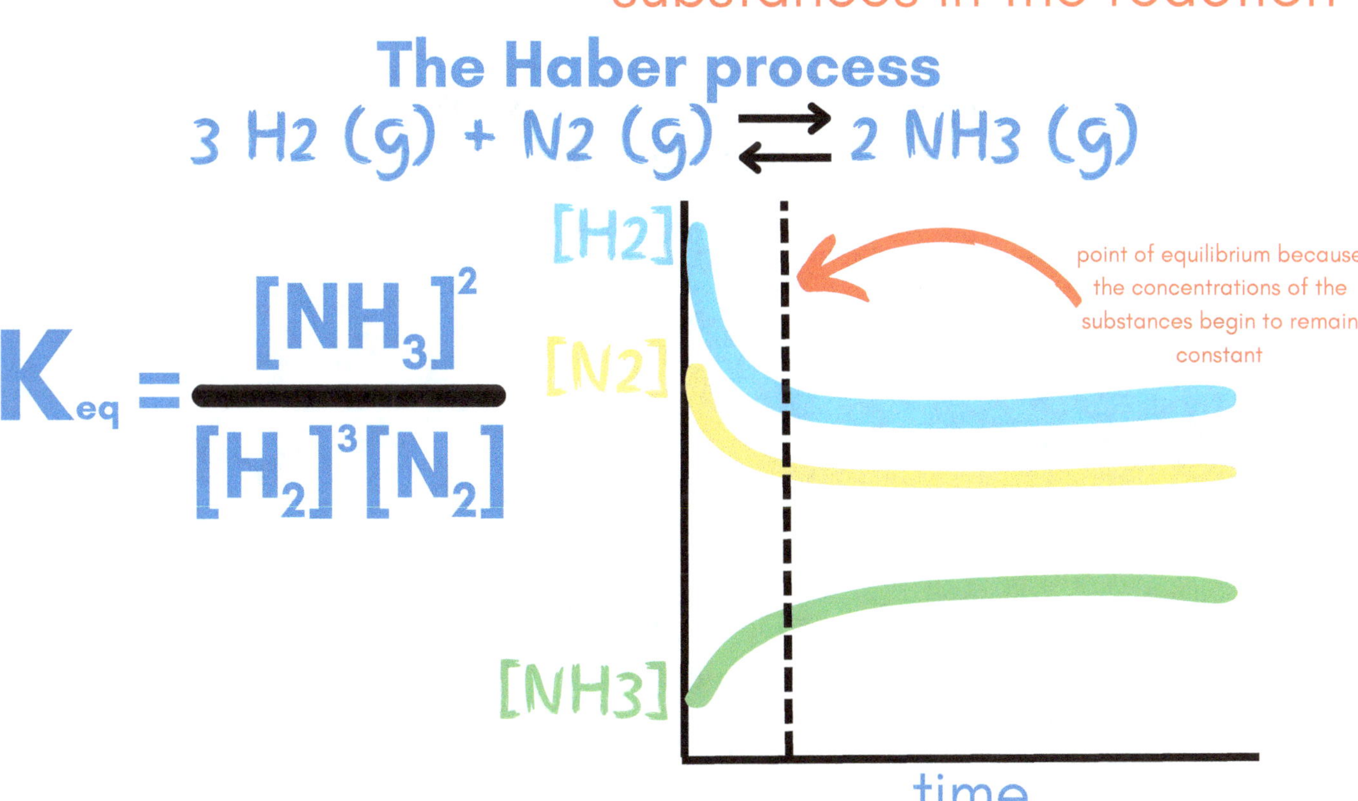

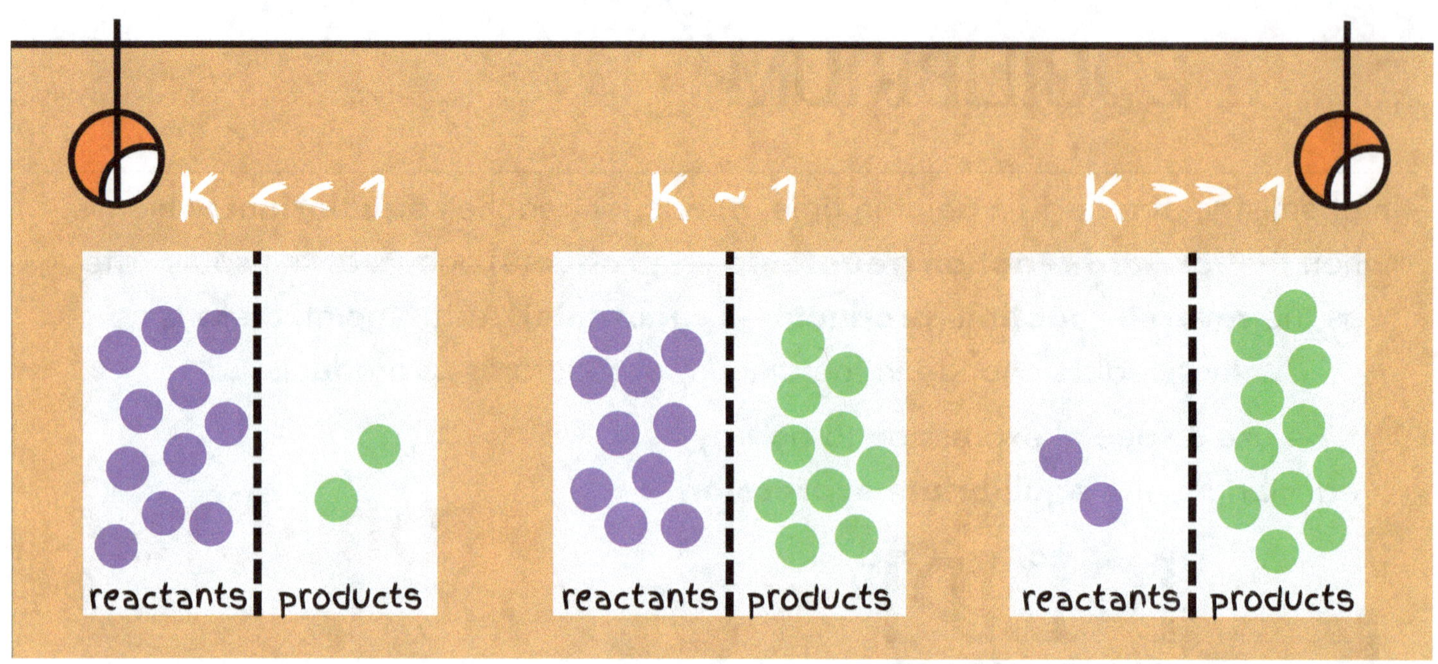

Just like **Hess's Law** in thermodynamics, there is a version of manipulating **K-values** in chemical equilibrium.

For **flipping** a reaction, instead of **negating** the **K-value,** we take the **reciprocal**.

For **increasing** or **decreasing** a reaction by a factor, we take the K-value to the **power** of that factor.

For **adding** reactions, instead of adding the K-values, we **multiply** the K-values together.

Using our knowledge of **K-values**, we can apply **Le-Chatelier's Principle** to determine which way our reaction will proceed to reach equilibrium if a stressor is introduced.

BASIC RULES

If we **increase** the **concentration** of reactants, the reaction will shift to the **right**. If we **decrease** the **concentration** of reactants, the reaction will shift to the **left**.

If we **increase** the **pressure** of gaseous reactants, the reaction will shift to the **right**. If we **decrease** the **pressure** of gaseous reactants, the reaction will shift to the **left**.

In an **exothermic reaction**, **increasing** the temperature causes the reaction to shift to the **left**, and **decreasing** the temperature causes the reaction to shift to the **right**.

In an **endothermic reaction**, **increasing** the temperature causes the reaction to shift to the **right**, and **decreasing** the temperature causes the reaction to shift to the **left**.

Adding a **catalyst** increases both forward and reverse reactions, so there is **no shift** in the reaction.

There is also a value for **any non-equilibrium** point in time called the **reaction quotient (Q)**. The expression that represents the Q-value is identical to the expression that represents the K-value, except the concentrations (or pressures) represent the *initial* concentrations (or pressures).

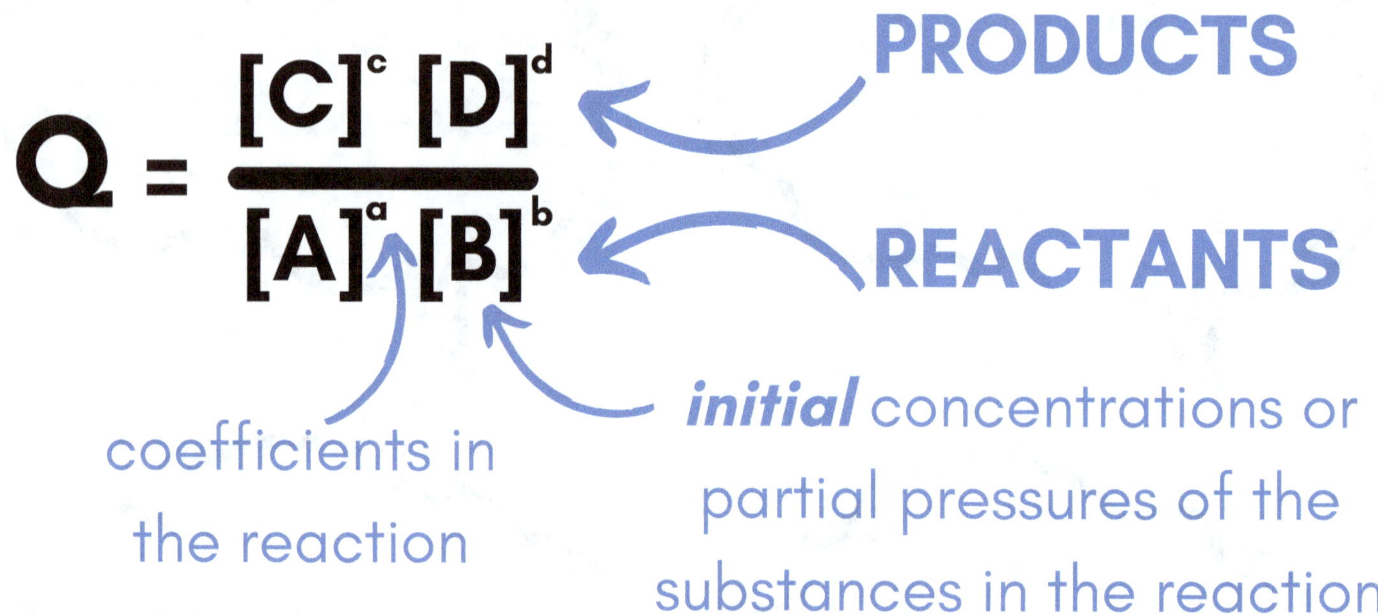

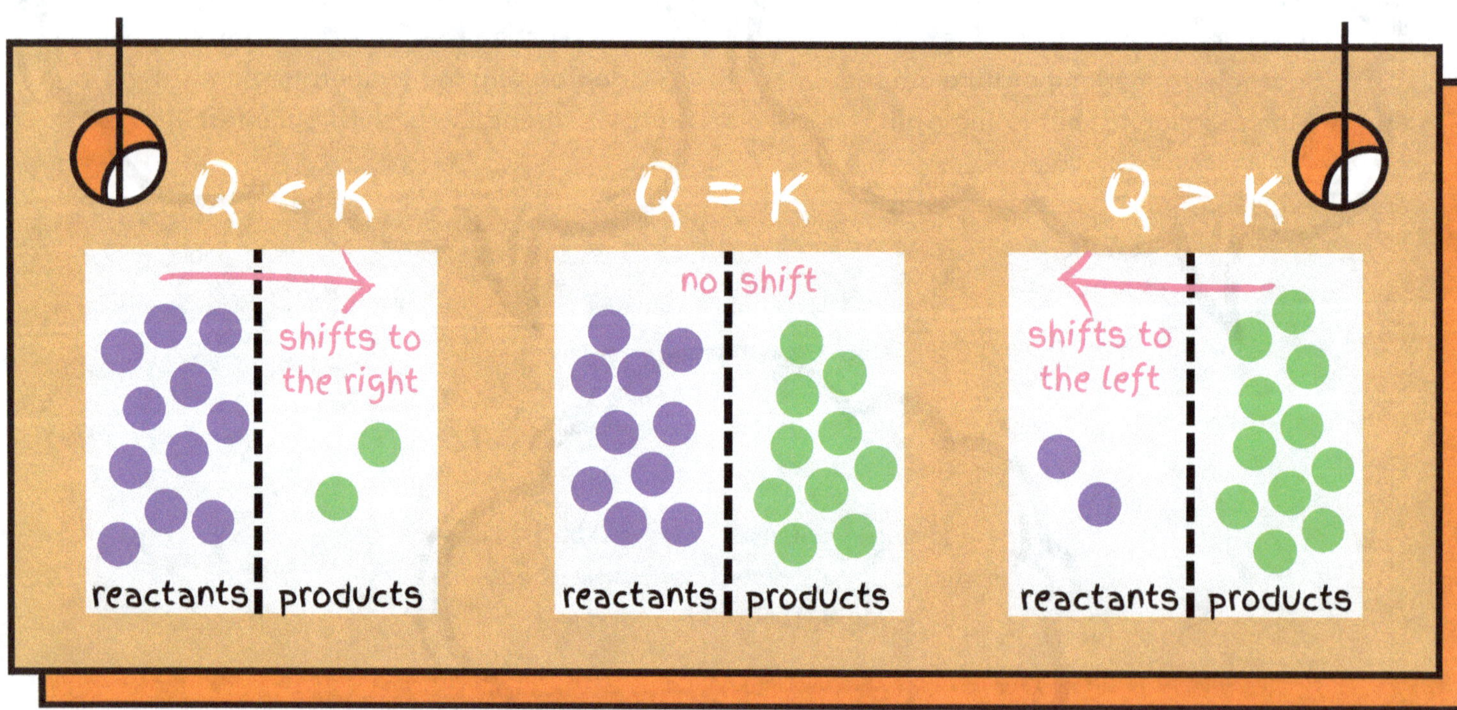

You may be tired of the new terms, but there is also a **solubility product** expression and various ways to solve problems involving aqueous solutions.

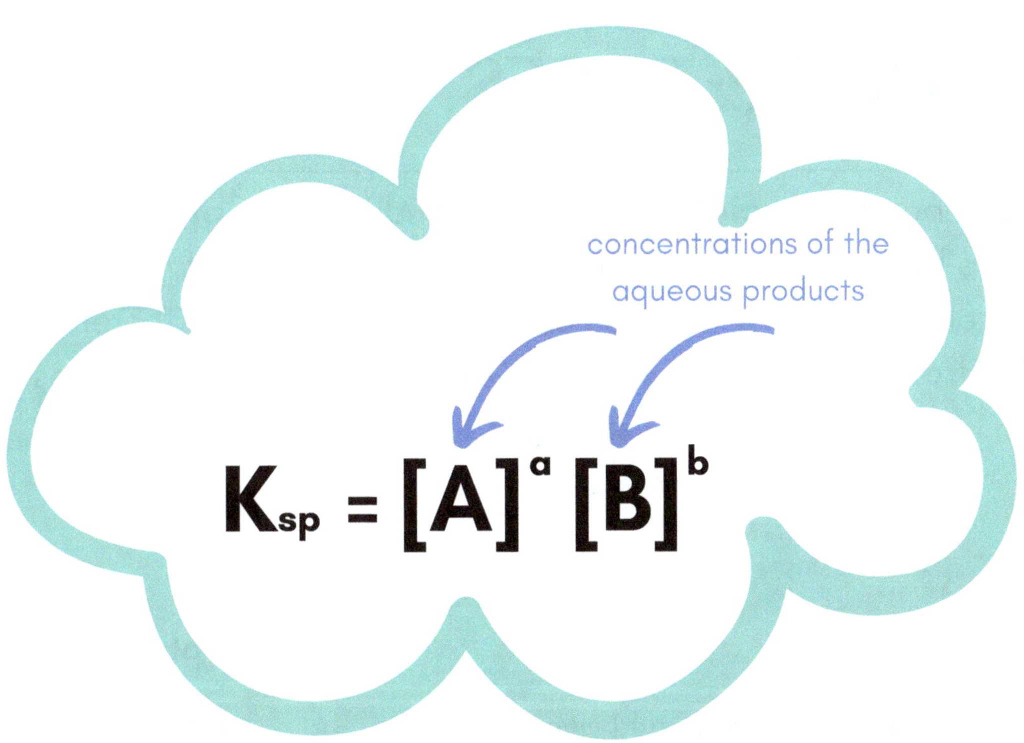

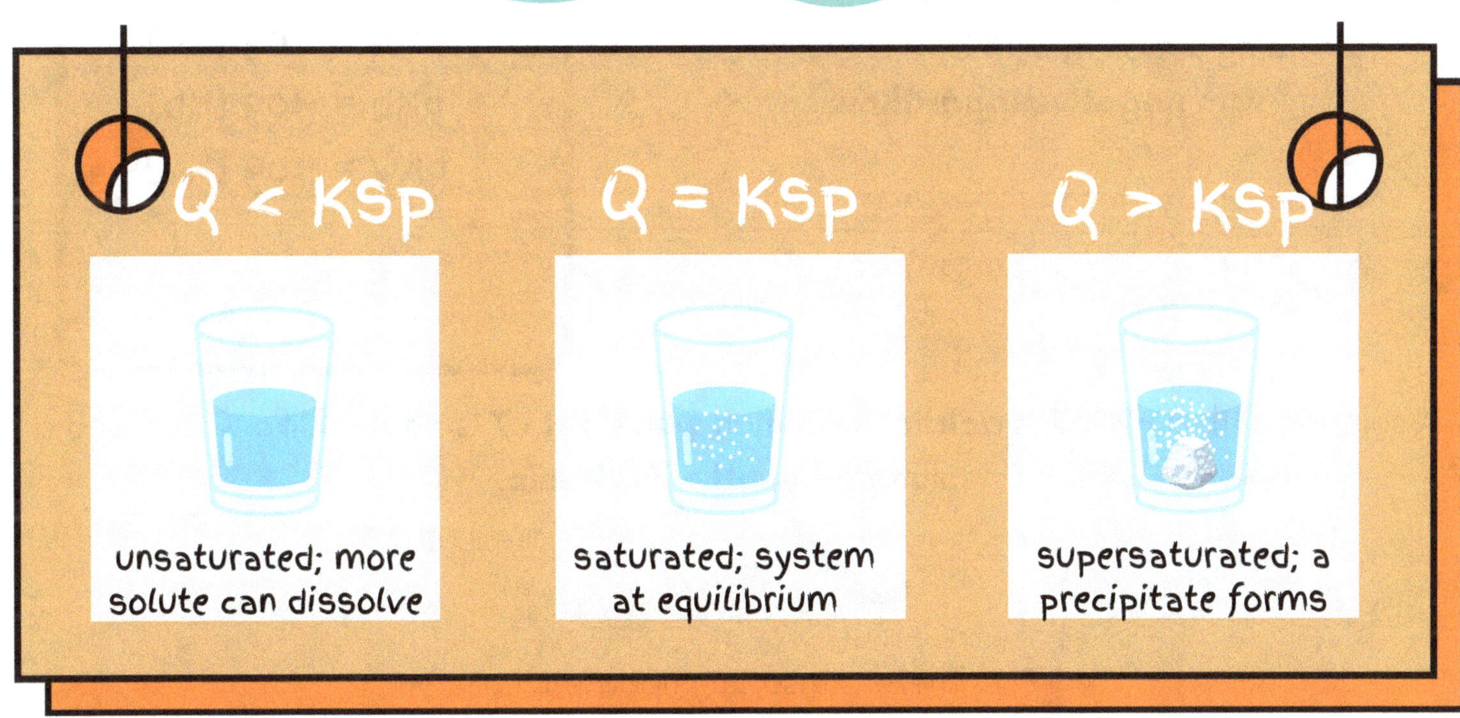

There is a phenomenon known as the **common-ion effect**. Essentially, if you have a solution of aqueous ions, to which you add a sample of an identical type of aqueous ions, the excess ions will **precipitate out**. For example, if you have a solution of **AgCl** and add **KCl**, the excess chloride ions will affect the equilibrium. Since **AgCl** is the **less soluble salt**, the excess chloride ions will cause **AgCl** to precipitate out.

To solve problems related to solubility, we use **ICE tables (Initial, Change, Equilibrium)**. We won't get that to this lesson, but just understand that it is a table that accounts for initial concentrations of ions, the amounts of ions that were added or used, and the end concentrations of ions.

A solution can be either **acidic** or **basic** depending on its **pH** and **pOH**. **pH** means "power of the hydrogen ion" whereas **pOH** means "*power of the hydroxide ion*." In acid-base scenarios, **hydrogen ions** decrease the pH, which thereby increases the pOH. **Hydroxide ions** increase the pOH, which thereby decreases the pH.

$$pH + pOH = 14.$$

pH, pOH, pK_a, pK_b, pK_w all have something in common: they all have to do with **negative logarithms**.

$$pH = -\log [H+]$$
$$pOH = -\log [OH-]$$
$$pKa = -\log [Ka]$$
$$pKb = -\log [Kb]$$
$$pKw = -\log [Kw]$$

A solution with a **pH < 7** is **acidic**, a solution with a **pH = 7** is **neutral** like water, and a solution with **pH > 7** is **basic**.

There is also an important term known as **acid strength**, which depends on many factors and determines how well an acid dissociates into hydrogen ions and other ions.

Strong acids dissociate completely in water, which means that the reaction goes to completion. **Weak acids** don't dissociate as well, so there is an **equilibrium** that occurs between the weak acid and water. Additionally, there is an important term known as **conjugates**, which are pairs of acids with their hydrogen ions and acids without their hydrogen ions. For example, **HCl** and **Cl**$^-$ are conjugates. In this case, HCl is the conjugate acid of Cl$^-$, and Cl$^-$ is the conjugate base of HCl. One important thing to understand is that **strong acids/bases** have conjugates that are **extremely weak**, and **weak acids/bases** have conjugates that are **weak**.

<u>Strong Acids & Bases</u>

ACIDS | BASES

ACIDS	BASES
HI	NaOH
HCl	LiOH
HBr	KOH
HNO3	Ca(OH)2
H2SO4	Sr(OH)2
HClO4	Ba(OH)2
HClO3	

Any other acid or base other than the ones listed above is most likely weak.

Strong acids dissociate into hydrogen ions and the resulting anions. For example, **HBr** dissociates into a hydrogen ion and a bromide ion. There are **three factors** that affect how strong an acid is: **bond polarity, bond strength,** and **conjugate base stability**.

Polarity occurs when one atom pulls electrons from a covalent bond toward itself. Therefore, if a bond is **more polar**, the electrons are pulled **away** from the hydrogen ion and toward the **anion**, which means that **as polarity increases, acid strength increases** since the hydrogen ion is able to come off more easily.

If a bond is strong, that means that the bond is harder to break. As a result, if the bond between the hydrogen ion and anion in acid is **strong**, the hydrogen ion will not be able to easily come off, making the acid **weaker**. Therefore, as **bond strength increases, acid strength decreases**.

Conjugate base stability refers to how stable the **anion** of an acid is without the hydrogen ion attached to it. If the anion is stable without the hydrogen ion, the acid will be **more likely** to donate the hydrogen ion, making the acid **stronger**. Therefore, **as conjugate base stability increases, acid strength increases**.

Acid Dissociation Constant

Base Dissociation Constant

$$K_a = \frac{[H^+][A^-]}{[HA]}$$

$$K_b = \frac{[HB^+][OH^-]}{[B]}$$

[H+] --> concentration of hydrogen ions

[A-] --> concentration of anions

[HA] --> concentration of acid

[HB+] --> concentration of conjugate acid

[OH-] --> concentration of hydroxide ions

[B] --> concentration of base

On a different note, **polyprotic acids** are acids that have multiple hydrogen ions that they can donate. An important thing to realize is that after the first hydrogen of a polyprotic acid is dissociated, the other hydrogen ions are **more difficult to dissociate**.

In the case where you have an **unknown concentration** of acid or base, you can use an **acid-base titration** to determine the concentration. In this case, you would add a **color-changing indicator** as well as a certain amount of either an **acid** or **base** of a **known concentration**, which will be called the **titrant**, to the **acid** or **base** of an **unknown concentration**, which will be called the **analyte**, to an **equivalence point**, which is when the indicator causes the entire mixture to **abruptly change color**. At this point, there will be an equal number of moles of both acid and base.

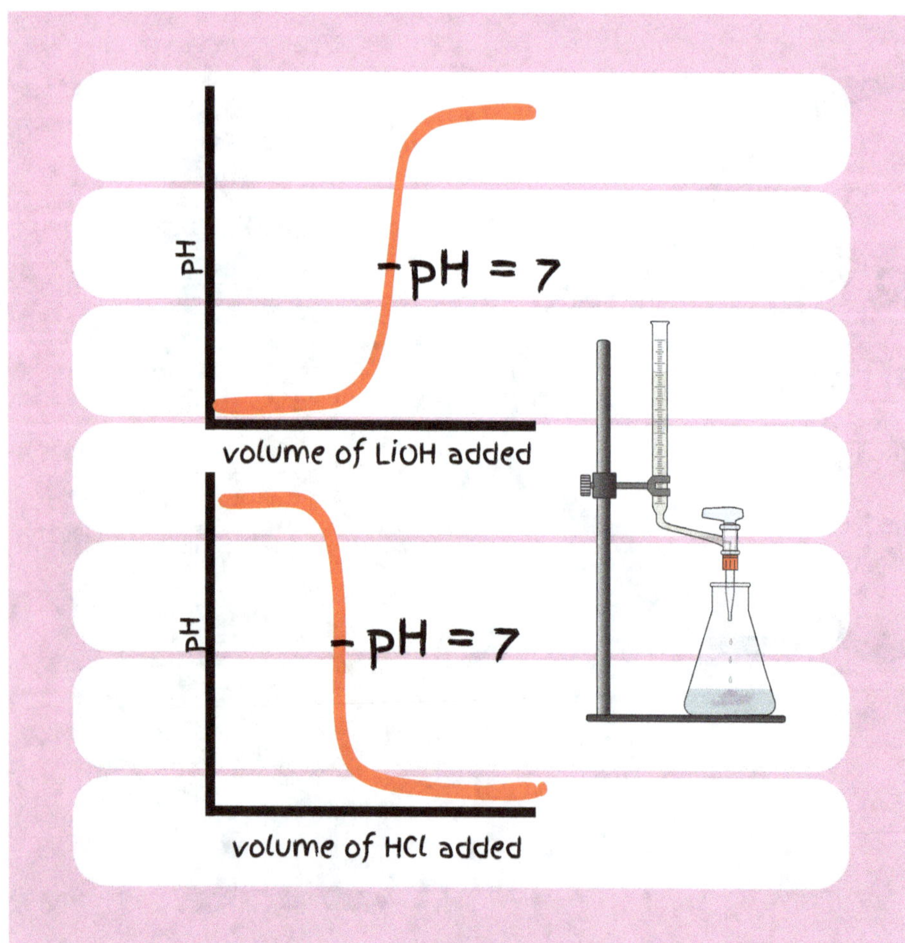

For a **strong acid + strong base** titration, the equivalence point is characterized by having a **pH of 7**. An example of such a titration is **titrating 25 mL of 0.10 M HCl** with **50 mL** of an unknown concentration of **LiOH** or **titrating 25 mL of 0.10 M NaOH** with **50 mL** of an unknown concentration of **HCl**. The graphs of the two scenarios are **opposites**.

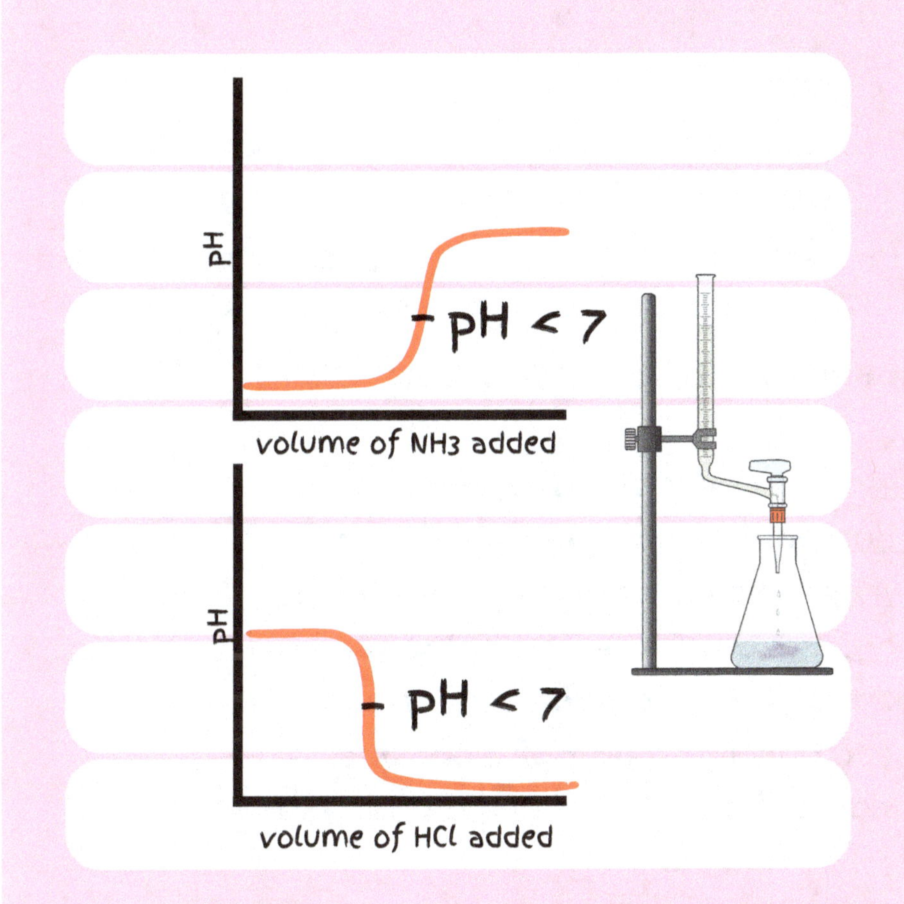

For a **strong acid + weak base** titration, the equivalence point is characterized by having a **pH of less than 7**. An example of such a titration is **titrating 25 mL of 0.10 M HCl** with **50 mL** of an unknown concentration of **NH_3** or **titrating 25 mL of 0.10 M NH_3** with **50 mL** of an unknown concentration of **HCl**.

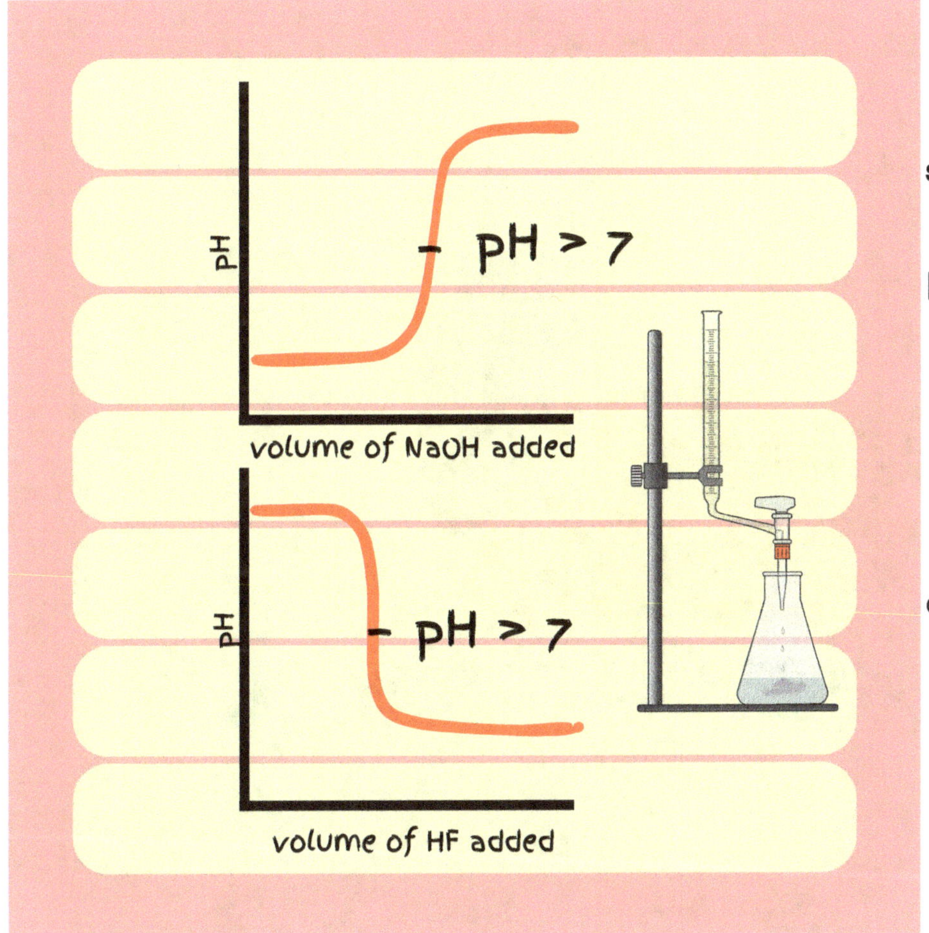

Lastly, for a **weak acid + strong base** titration, the equivalence point has a **pH of greater than 7**. An example of such a titration is **titrating 25 mL of 0.10 M HF** with **50 mL** of an unknown concentration of **NaOH** or **titrating 25 mL of 0.10 M NaOH** with **50 mL** of an unknown concentration of **HF**.

Vocabulary That We Didn't Cover:

<u>Arrhenius Acid</u>: a donor of a **hydrogen ion**.
<u>Arrhenius Base</u>: a donor of a **hydroxide ion**.
<u>Brönsted-Lowry Acid</u>: a donor of a **hydrogen ion**.
<u>Brönsted-Lowry Base</u>: a receiver of a **hydrogen ion**.
<u>Lewis Acid</u>: a receiver of a **pair of electrons**.
<u>Lewis Base</u>: a donor of a **pair of electrons**.
<u>Amphoteric</u>: can be both an acid and a base, like water.
<u>Buffer</u>: can be mixed into a solution to **prevent** a lot of pH change.

11 FREE ENERGY & SPONTANEITY

Entropy is how scientists can determine how **chaotic** the world is. It is a measure of **disorder**, which depends on **temperature**, **energy**, and **state of matter**. Entropy is represented by a capital **S**.

Since the actual entropy value of a reaction is difficult to calculate, we rely on the *change* in entropy.

The **change in entropy** is determined by subtracting the **entropy of reactants** from the **entropy of products**. To **increase** the entropy (or disorder) of a system, a reaction can convert a **solid to liquid** or **liquid to gas**, produce **more aqueous ions** or **gas molecules**, increase the **temperature**, increase the **volume** of the system, etc. Essentially, anything that can make a reaction more "chaotic" results in a **positive entropy change**, which means that entropy increases as the reaction proceeds.

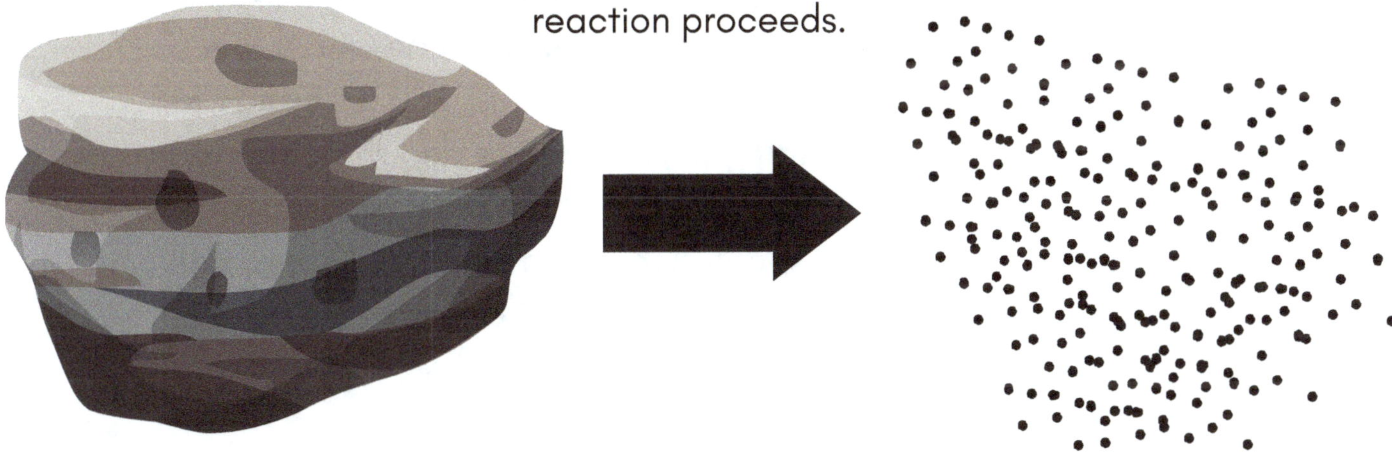

Bridging off from entropy, there is also a concept known as **Gibbs free energy**, which is a measure of the energy available to trigger a reaction. It can be **positive** or **negative**, and depending on the sign of the Gibbs free energy, the reaction can be **spontaneous** or **non-spontaneous.** We also measure Gibbs free energy as the **change** in Gibbs free energy.

If a reaction is **spontaneous**, it can occur of its own accord without energy. If a reaction is **non-spontaneous**, it requires energy to occur.

There are **four scenarios** that determine the **spontaneity** of a reaction:

If the **change in enthalpy (H)** is **negative** and the **change in entropy (S)** is **positive**, then the **Gibbs free energy (G)** will be **negative**, and the reaction will **always be spontaneous**. If the **change in enthalpy (H)** is **positive** and the **change in entropy (S)** is **negative**, then the **Gibbs free energy (G)** will be **positive**, and the reaction will **never be spontaneous**.

Secondly, if the **change in enthalpy (H)** is **negative** and the **change in entropy (S)** is **negative**, then the reaction will be spontaneous in a situation with a **low temperature**. However, if the **change in enthalpy (H)** is **positive** and the **change in entropy (S)** is **positive**, then the reaction will be spontaneous in a situation with a **high temperature**.

There are also two important equations on the next page!

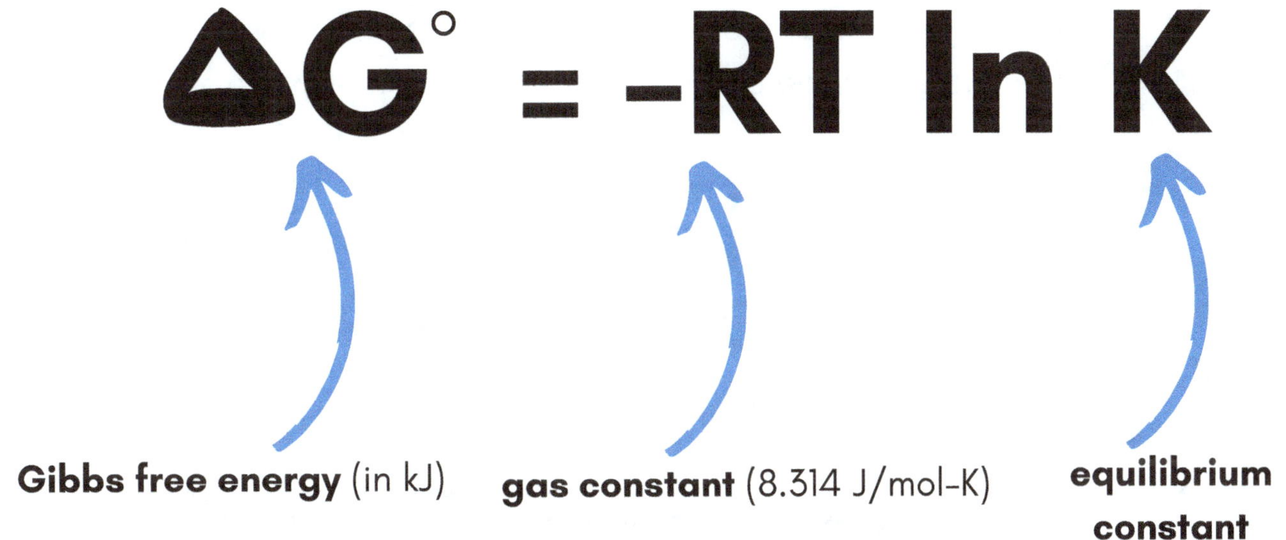

Use these equations to solve problems related to spontaneity, entropy, and Gibbs free energy!

12 ELECTROCHEMISTRY 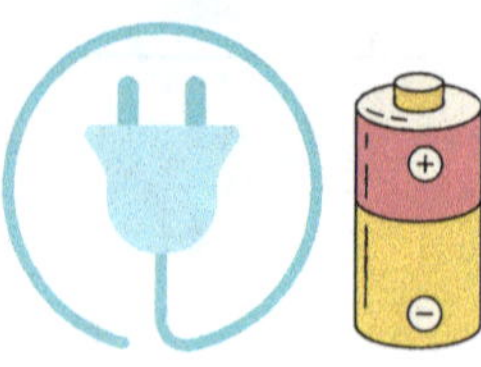

Electrochemistry is the study of how **electricity** works in terms of differences in electrical charge. It encompasses many topics, including **redox reactions** and **battery usage**. Before moving on to actual electrochemistry, you should understand what a **redox reaction** is.

In a **redox reaction**, as stated earlier, there are **two parts**: **oxidation** and **reduction**. **Oxidation** occurs when a species **loses electrons** and becomes more **positive**, and **reduction** occurs when a species **gains electrons** and becomes more **negative**. In this manner, the **overall electrical charge** stays constant as there is a species that becomes more positive and another that becomes more negative.

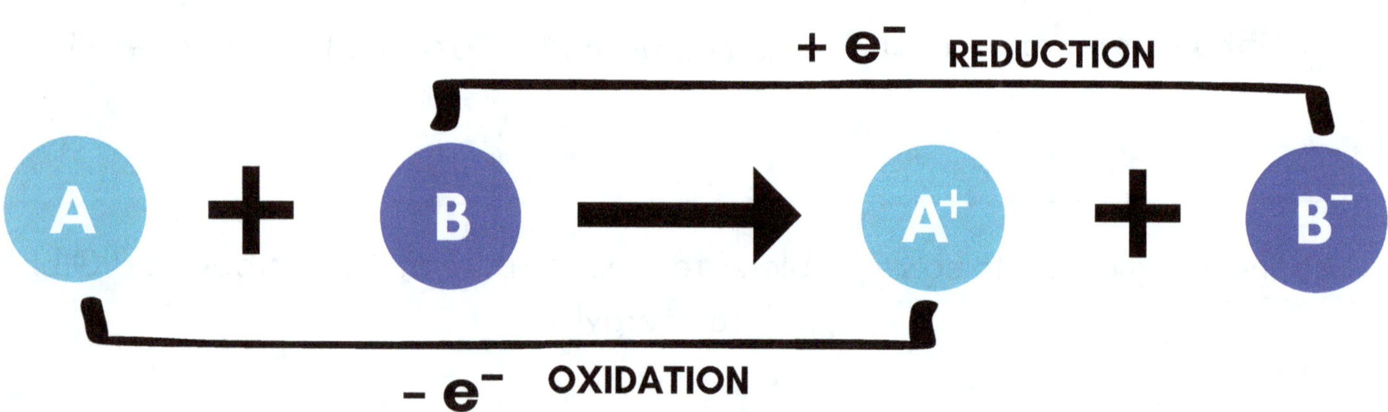

In a compound or polyatomic ion, each element has an **oxidation number**, which tells us how many electrons that element must lose to form a bond with another species. To determine the oxidation number, you should take the **overall charge** on the molecule into account, as all of the oxidation numbers for the elements should add up to the overall charge. Take **$KMnO_4$**, for example.

The total charge on the compound is **zero**, which means that the oxidation numbers for **potassium, manganese,** and **oxygen** should add up to **zero.**

$$KMnO_4$$

Now, we can separate the compound into **K^+** and **MnO_4^-**, the latter of which is one of the polyatomic ions that you should memorize.

$$K^+ \qquad MnO_4^-$$

Any ion formed from **Group 1 elements** have a +1 oxidation number, which means that the **potassium ion** should have a **+1** oxidation number, making the **MnO_4^-** have a **-1** oxidation number.

$$MnO_4^-$$

Unless **oxygen** is in **hydrogen peroxide (H_2O_2)**, which would make the oxygen have a **-1** oxidation number or combined with **fluorine**, which would make the **oxygen** have a **positive** oxidation number, oxygen always has a **-2** oxidation number. Using a bit of algebra, you can figure out the oxidation number of **manganese**, which is **+7**.

$$MnO_4^-$$

$$X + 4(-2) = -1$$
$$X - 8 = -1, \; X = +7$$

$$KMnO_4$$
$$(+1) + (+7) + 4(-2) = 0$$

There are two concepts: **reduction potential** and **oxidation potential**. The **reduction potential** is how likely an atom is to be reduced whereas the **oxidation potential** is how likely an atom is to be oxidized. We focus more on **reduction potentials**, which are listed on a **chart**. **Reduction potentials (E)** can be **positive** and **negative**, and a **positive** reduction potential indicates that the atom is likely to be **reduced** whereas a **negative** reduction potential indicates that the atom is likely to be **oxidized**.

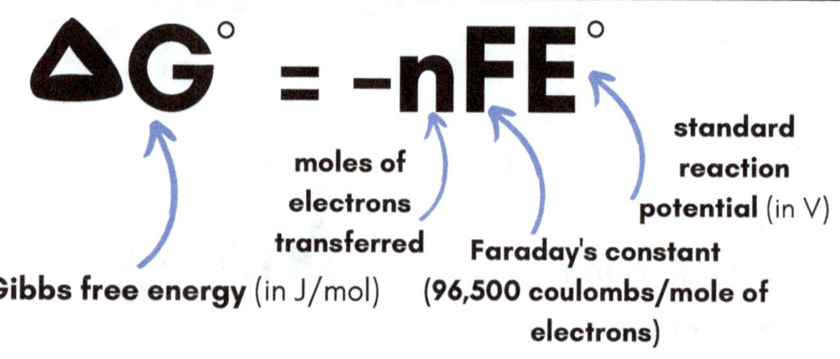

Take a look at the following example:

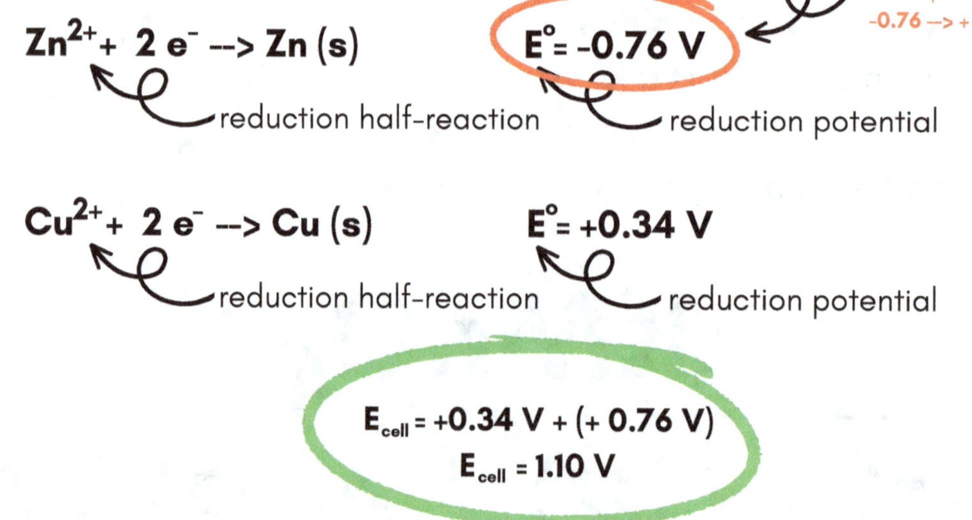

In a **galvanic** (aka **voltaic**) **cell**, which utilizes **chemical energy** to produce **electrical energy**, a **spontaneous reaction** occurs, and the **two half-reactions** - **oxidation** and **reduction** - take place in **separate** containers housing aqueous solutions of metal ions. There is a **transfer of electrons** from the **anode** (where **oxidation** occurs) to the **cathode** (where **reduction** occurs), and the electrons pass through a **wire** connecting the two **electrodes** (a generalization of anode and cathode). As the **anode** gets more **positive** from losing electrons and the **cathode** gets more **negative** from gaining electrons, the **salt bridge** maintains electrical **neutrality** as negative ions go into the anode and positive ions go into the cathode.

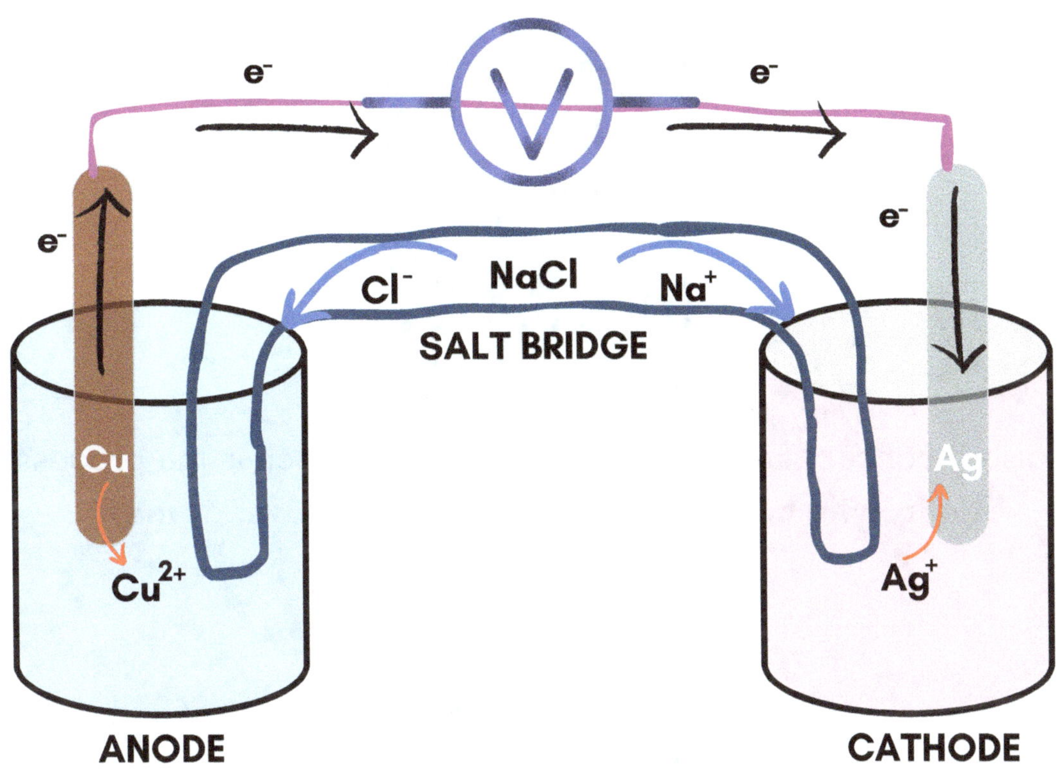

The metal at the anode gets oxidized, and the electrons lost enter the wire, and the resulting cation go into the solution.

The metal at the cathode gets reduced by the electrons from the wire, and the solid metal builds up on the cathode metal bar.

An **electrolytic cell**, on the other hand, causes a **non-spontaneous reaction** to occur by supplying energy. An outside source of **voltage** is used to allow an **unfavorable redox reaction** to occur. A **battery** is used to power the reaction, and the **anode** usually produces **gas** whereas the **cathode** usually produces **metal**. The below diagram is a **basic diagram** of the electrolytic cell of a **molten salt**.

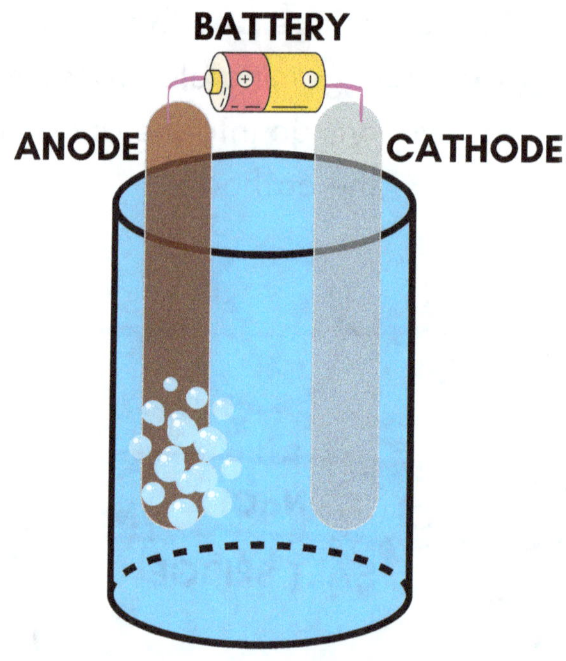

There is also a concept known as **electroplating**, which is the process of using **electrolytic cells** to cover an object in a layer of **metal**.

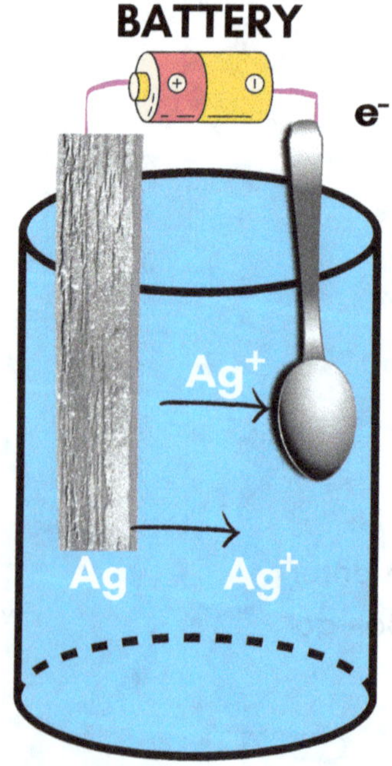

You can use **mathematical equations** to solve problems regarding electroplating!

And **THAT'S A WRAP!**
You just learned a **huge chunk** of chemistry with this novel!
I hope you now understand the fundamentals of chemistry and how important chemistry is to our lives today.

See you soon! :)